LIBRARY IN A BOOK

NUCLEAR POWER

David E. Newton

Facts On File

An imprint of Infobase Publishing

For Eileen and Ashley
More than just family . . . Our friends!

NUCLEAR POWER

Copyright © 2006 by David E. Newton
Map and graphs copyright © 2006 by Infobase Publishing

Facts On File, Inc.
An imprint of Infobase Publishing
132 West 31st Street
New York NY 10001

Library of Congress Cataloging-in-Publication Data
Newton, David E.
    Nuclear power / David E. Newton.
        p. cm.—(Library in a book)
    Includes bibliographical references and index.
    ISBN 0-8160-5655-2
  1. Nuclear engineering. I Title. II. Series.
    TK9146.N453 2005
    333.792′4—dc22   2005006437

Text design by Ron Monteleone
Map and graphs by Sholto Ainslie

Printed in the United States of America

MP Hermitage 10 9 8 7 6 5 4 3 2 1

This book is printed on acid-free paper.

CONTENTS

PART III
APPENDICES

PART I

OVERVIEW OF THE TOPIC

CHAPTER 1

AN INTRODUCTION TO NUCLEAR POWER AND ISSUES OVER ITS USE IN THE UNITED STATES

On August 6, 1945, a B-29 bomber from the U.S. Air Force, the *Enola Gay*, dropped a single bomb on the Japanese city of Hiroshima. That act was a desperate effort by the U.S. military to bring World War II—which had stretched on for nearly four years—to a quick and decisive conclusion. The bomb carried by the *Enola Gay* was a new kind of explosive device, a *nuclear* (or *atomic*) weapon. It had been given the nickname Little Boy and measured about three meters (10 feet) in length and 0.7 meters (two feet) in diameter. Little Boy was the most destructive weapon ever produced by humans, with a power equivalent to that of about 15,000 tons of TNT, the most powerful conventional explosive known. For this reason, Little Boy was classified as a 15-kiloton (or 15-kt) bomb.

The *Enola Gay's* mission was a military success and a civilian and humanitarian disaster. Official estimates placed the number of individuals killed at 118,661, with another 79,130 persons wounded and 3,677 missing. An area of about 13 square kilometers (five square miles) was essentially reduced to rubble, with the remnants of only a single building, the Hiroshima Prefectural Industrial Promotion Hall, left standing at Ground Zero (the point beneath which the bomb was detonated). The effects of Little Boy and a similar bomb (Fat Man) dropped on Nagasaki three days later convinced the Japanese government of the futility of continuing its war efforts, and it sued for peace a day later. The bombing of Hiroshima and Nagasaki is often cited as the beginning of the Nuclear Age. But was it really?

Certainly, the events of August 6 and 9, 1945, brought nuclear power to the attention of the world in a sudden and dramatic way. The concept of

3

splitting atoms to produce energy (hence, the term most widely used then: *atomic power*) was essentially unknown to virtually all scientists and nonscientists alike at the time. But that concept had its roots in scientific principles going back well over a hundred years. And efforts at finding a way of unleashing the enormous amounts of energy stored within the atomic nucleus had been going on, in one form or another, for nearly a decade.

The existence of nuclear energy (as it is more properly called) had been intuitively obvious to scientists since the discovery in the early 20th century of the proton by English physicist Lord Ernest Rutherford (1871–1937). Rutherford demonstrated that the central core of nearly all atoms—the nucleus—contains two or more protons, often large numbers of protons. Protons all carry the same charge, a single unit of positive electricity. Scientists have also known that particles with like charges tend to repel each other strongly. Even more important, the closer such particles are to each other, the stronger the force of repulsion. So, how is it possible that two or more (often, many more) like-charged particles can huddle together within an atomic nucleus?

The only answer one can imagine is that some very strong force (or forces) must exist within the atomic nucleus that holds these like-charged protons together. After all, most atomic nuclei found in nature are stable. That is, they do not fly apart, as one might expect from a closely packed assemblage of like-charged particles. So, even without the vaguest knowledge as to what these nuclear forces might *be*, scientists were fairly certain they must exist.

THE SCIENCE BEHIND NUCLEAR ENERGY

The real beginning of the age of nuclear power can, perhaps, be traced to the discovery of the atom by English chemist and physicist John Dalton (1766–1844) in 1803. Dalton did not so much "discover" the atom, in terms of "finding" or "seeing" it, as he did explain why particles such as atoms *had* to exist and why such particles could explain so many chemical phenomena then known. The atoms that Dalton envisioned were very small, solid, indivisible particles, somewhat like extremely tiny marbles or ball bearings.

For nearly a century, Dalton's image of atoms served scientists well. But toward the end of the 19th century, new information became available showing that Dalton's view of atoms was overly simplified and incomplete. The key discovery during this period was that of English physicist J. J. Thomson (1856–1940). In 1897, Thomson found that atoms are not solid and indivisible but instead consist of at least two parts. One part is a negatively charged particle discovered by Thomson, which was later given the name *electron*. The second part of the atom was a still-unidentified, amorphous mass that was later given the name nucleus.

Thomson's discovery of the electron began a process of dissecting the atom, a process that might be compared to the dissection of an apple. That process eventually led to the discovery of two particles that make up an atomic nucleus. One of the two particles, the proton, was found by Rutherford between 1911 and 1914. A proton is a hydrogen atom without its electron. Then, in 1932, the second nuclear particle, the neutron, was discovered by English physicist James Chadwick (1891–1974). As a result of this long line of research, scientists now have a very good understanding not only of the particles that make up the atom but also of the forces that hold those particles together within an atom.

TRANSMUTING ELEMENTS

The process of collecting this information has not been an easy one. Obviously, one cannot cut open an atom or an atomic nucleus, no matter how small a knife is available, the way one cuts open an apple. One widely popular method for studying the structure of an atom and nucleus has long been to bombard them with beams of energy and/or particles. By the early 20th century, scientists had a number of such tools available to them: X-rays, discovered by French physicist Antoine Henri Becquerel (1852–1908) in 1896; gamma rays, discovered by French physicist Paul Villard (1860–1934) in 1900; and alpha and beta rays, both discovered by Rutherford in 1897.

In this form of analysis, a beam of energy is made to collide with an atom (actually, a very large group of identical atoms). The products of that collision are then analyzed to see what atoms and other particles are present. The first experiment of this kind was designed and carried out, once again, by Lord Rutherford. In this experiment, nitrogen gas was exposed to a beam of alpha particles (which constitute an alpha ray). When Rutherford analyzed the products of this reaction, he found they consisted of oxygen and hydrogen atoms.

This experiment was important for two reasons. First, Rutherford concluded that the hydrogen gas formed in the reaction must have come from protons expelled from the nucleus of nitrogen atoms. This result convinced him (and other scientists) that protons are one of the constituent particles that make up atomic nuclei. Second, the experiment showed that atoms can be transmuted, that is, changed from one form into another form. In this case, nitrogen was transmuted, or changed, into oxygen.

By the early 1930s, scientists throughout the world were studying every possible permutation of this kind of experiment, bombarding virtually every kind of atom with every form of energy available. In one such experiment, Chadwick bombarded beryllium atoms with alpha rays. He observed the formation not only of carbon but also of a new kind of particle with no electrical charge but with a mass of one, a particle to which he gave the name neutron. (The term *mass* corresponds closely to the more common word weight.)

One of the most common variations of these bombardment experiments was called an *n, γ reaction*. The name comes from the fact that some type of element is bombarded with neutrons (whose symbol is *n*), resulting in the formation of gamma rays (whose symbol is *γ*) as one product of the reaction.

The interesting feature of an n,γ reaction is that it generally results in the formation of an element one position higher in the periodic table than the original (target) element. The periodic table is a chart that contains all the chemical elements arranged in sequence, according to their increasing size. For example, bombarding sodium atoms with neutrons results in the formation of calcium atoms, one position higher in the periodic table. Similarly, if calcium atoms are bombarded with neutrons, the next heavier element, aluminum, is formed. And, neutron bombardment of aluminum results in the formation of silicon, the next heavier element in the periodic table.

Scientists in the 1930s had a field day with n,γ reactions, at least partly because they represented the fulfillment of one of the oldest goals of chemical science: the transmutation of matter. As far back as ancient Egypt, certain scholars known as alchemists had been searching for ways to change "base" (common) metals, such as lead and iron, into "noble" (valuable) metals, such as silver and gold. Stories of these efforts, along with the political intrigues they inspired, would fill many books the size of this one. Despite their efforts, though, no one was successful in transmuting an element until the 1930s, when untold numbers of scientists were accomplishing the act with relative ease (though none were able to turn lead into gold).

One of the reactions of special interest during this period involved the bombardment of the element uranium with neutrons. An n,γ reaction with uranium would be expected to result in the formation of the next heavier element in the periodic table. Uranium, however, is the heaviest element in nature; it is the last naturally occurring element in the periodic table. The bombardment of uranium with neutrons, if a reaction could occur, would result therefore in the formation of a new element, an element not found in nature.

One of the research teams most actively studying n,γ reactions in the early 1930s was led by Italian physicist Enrico Fermi (1901–54). In a series of experiments conducted in 1934, Fermi's team bombarded uranium with neutrons and became convinced that they had produced a new element—in fact two new elements—with atomic numbers greater than that of uranium. The Fermi team named the elements *auserium* and *hesperium* in honor of ancient names of Italy. Fermi announced the discovery of these elements at his Nobel Prize lecture in 1938. Other scientists were unable to replicate Fermi's results, however, and his claim for two new elements had to be rejected. However, only a few years later, it became obvious that Fermi's team had realized an accomplishment at least as important as that of transmutation. They had actually succeeded in splitting apart uranium atoms, an entirely new type of reaction that had not even been imagined by Fermi, his research team, or his colleagues in the community of physicists. This process would become known as fission.

THE DISCOVERY OF NUCLEAR ENERGY

Enrico Fermi was by no means the only scientist interested in the effects of bombarding uranium with neutrons. In their laboratory at the Kaiser Wilhelm Institute in Berlin, German chemists Otto Hahn (1879–1968) and Fritz Strassman (1902–80) carried out similar experiments. When Hahn and Strassman analyzed the result of their bombardment of uranium with neutrons, they found what appeared to be evidence that the elements barium and krypton had been produced in the reaction. Barium and krypton are elements whose atoms are each about half the size of a uranium atom. Such results seemed impossible. They would mean that the bombardment of uranium with neutrons had not produced the next heavier element (as was typically the case with n,γ reactions, including Fermi's) but the formation of two elements, each about half the size of uranium. That is, the uranium atom appeared to have been split into two roughly equal parts, a barium atom and a krypton atom.

Hahn and Strassman found it difficult to accept the apparent explanation of their experiment. Those results were unprecedented and seemingly uninterpretable by any existing physical theory. They decided to send their data to Lise Meitner (1878–1968), an Austrian physicist and former colleague of Hahn then living in Sweden. Meitner had fled from Germany in 1938 to escape the Nazi purge of Jewish scientists. Meitner and her nephew Otto Frisch (1904–79) derived a physical and mathematical explanation of the Hahn-Strassman results. They showed that the reaction observed by Hahn and Strassman had indeed occurred and that neutrons had brought about fission (splitting) of the uranium nucleus.

The process of atomic fission posed a number of new problems for scientists. One of the most intriguing was the question of the "lost neutrons" in the reaction. It is easy enough to count up the number of neutrons present in the original uranium atom (there are 146) and the number of neutrons present in the barium and krypton atoms formed in the reaction (the number varies, but it is always less than 146). What happened to the uranium neutrons that are not present in the barium and krypton atoms? They could not simply "lose" particles of matter!

The explanation for this puzzle is actually quite simple. During the process of atomic fission, some neutrons from the uranium nucleus are set free. They are not used to build nuclei of new atoms, such as barium and krypton. They simply escape into the surrounding environment.

The existence of these "leftover" neutrons in a fission reaction was first discovered by Fermi in January 1939. The scientific world was immediately aware of the profound significance of this discovery. If neutrons are produced as the result of a reaction they initiate, then, once the reaction has begun, the reaction will continue on its own accord. Such reactions are known as chain reactions. The fissioning of a single uranium nucleus can release neutrons that can then be used to fission other uranium atoms, which can then fission and produce more neutrons, and so on. Such chain reactions continue, at least in theory, as long as uranium atoms are present to be fissioned.

A second puzzling feature of fission reactions might be called the "lost mass" problem. This problem was originally even more troubling than that of the "lost neutrons." It is possible to measure very accurately the mass of all the particles involved in a fission reaction. When that calculation is made, an interesting result is obtained. The total mass of all the particles present at the beginning of the reaction (a neutron and a uranium atom) is just slightly *greater than* the total mass of all the particles produced in the reaction (neutrons, barium nucleus, and krypton nucleus). The amount of mass that "disappears," the so-called mass defect, is very small, less than one-tenth of 1 percent of the total mass of the uranium atom. Nonetheless, some explanation must be found for this "lost" mass, just as it was necessary to explain the "lost" neutrons.

Surprisingly, that explanation is also fairly simple and had been known for more than two decades. In 1905, the great Austrian-American physicist Albert Einstein (1879–1955) announced his General Theory of Relativity, one component of which involved the equivalence of matter and energy. Einstein showed that a given amount of matter (m) has an energy equivalence (E) that can be calculated by means of a now-famous equation, $E = mc^2$, where c is the speed of light. That is, the energy equivalence of matter can be determined by multiplying the mass of that matter by the square of the speed of light (30 million meters per second). No matter how small a sample of matter selected, its energy equivalence will be very large because that mass is being multiplied by a very large number, 90,000,000,000,000,000 meters^2 per second^2. Each time a single uranium nucleus is fissioned, only a tiny amount of matter is lost, and a somewhat larger amount of energy is produced. But in a nuclear chain reaction, billions and billions of uranium atoms are fissioned in a fraction of a second. In that scenario, large amounts of energy are produced very rapidly.

THE DEVELOPMENT OF NUCLEAR WEAPONS

By early 1939, the practical significance of this scientific information had become obvious to a small group of scientists in various countries of the world. One of the first of these scientists to act on that information was Enrico Fermi, who in March 1939 gave a talk to a group of researchers at the U.S. Department of the Navy. Fermi outlined the potential application of nuclear energy for the development of an entirely new class of weapons, those based on nuclear chain reactions. The Navy Department showed little interest in the idea, however, and Fermi and his colleagues decided to be more aggressive in publicizing the news about nuclear chain reactions.

In summer 1939, two of Fermi's colleagues, Hungarian-American physicists Leo Szilard (1898–1964) and Eugene Wigner (1902–95), decided to write a letter to President Franklin D. Roosevelt outlining the potential military significance of nuclear chain reactions. They convinced the world's greatest living scientist, Albert Einstein, to sign the letter, of which there were actually two forms, a "short" form and a "long" form. The long form of the letter was delivered to Roosevelt on November 11, 1939. Convinced of the issue's significance, the president appointed a group of individuals, called the Uranium Committee, to study the military implications of nuclear energy. Only 10 days later, the committee held its first meeting.

Although the Uranium Committee somewhat reluctantly recommended a program of research on the development of nuclear weapons, its report soon disappeared into White House files, and virtually no action was taken. Later reports met with a similar fate, and it was not until October 1941 that President Roosevelt was finally convinced of the importance and urgency of a nuclear weapons project. The president's decision to act on nuclear weapons research was strongly influenced by a report written by a group of British scientists, the MAUD report, suggesting the likelihood that German scientists had already begun—or would soon initiate—such a program in their own country. President Roosevelt authorized the creation of a special research project, to be known as the Manhattan Engineer District, to explore the possibility of developing nuclear chain reactions for military purposes. The project eventually became better known as simply the Manhattan Project.

Nuclear Power

THE MANHATTAN PROJECT AND FISSION BOMBS

The task facing scientists working on the Manhattan Project was a daunting one. Endless numbers of questions in both basic and applied science had to be answered. Researchers from virtually every scientific discipline were assigned to a host of projects dealing with topics ranging from the separation of uranium isotopes for use in the bomb to the design of bomb detonation mechanisms. Isotopes are different forms of the same element with the same atomic number but different atomic masses.

Much of the earliest research was carried out at the University of Chicago's Metallurgical Laboratories (Met Lab), under the direction of American physicist Arthur Compton (1892–1962). Met Lab's first challenge was to prove that nuclear chain reactions really can occur. Theoretical calculations seemed to show conclusively that they could occur and that they would result in the release of huge amounts of energy, but no one had ever tested those theories with actual experiments, due to the dangerous risks involved.

The goal of researchers at Met Lab, then, was to produce a *controlled* nuclear chain reaction, a chain reaction that actually took place but without the release of very large amounts of energy. That is, their goal was to prove that chain reactions can occur without blowing up the University of Chicago and the surrounding city. In technical terms, this challenge meant achieving a nuclear chain reaction in which one neutron was produced for each neutron used up. Under those conditions, the reaction could continue on its own without further input from humans, but it would go slowly enough to release only a modest (safe) amount of energy.

That task was accomplished on December 2, 1942, about a year after the laboratory had begun operation. The working model developed by Met Lab researchers, called Chicago Pile One, had essentially the same composition as a present-day nuclear power plant reactor, the unit in a nuclear power plant where energy is actually generated. (The term *pile* is a somewhat obsolete term that refers to the collection of materials and devices within which a nuclear chain reaction occurs. The newer and generally synonymous term for "pile" is nuclear reactor.)

The Chicago One Pile contained the power source itself, a combination of uranium metal and uranium oxide; graphite, a material used to slow down neutrons; and control rods, cylinders containing cadmium metal. The function of the graphite was to slow down the speed of neutrons released during fission. Fission occurs only with neutrons moving at relatively slow speeds, so some mechanism for reducing their velocities is necessary. The cadmium present in the control rods has the property of absorbing neutrons and "taking them out of circulation" in the nuclear chain reaction. By raising or lowering the position of the control rods, operators could regulate the number of neutrons present in the pile and, hence, the rate at which the reaction proceeded.

Confirmation that a nuclear chain reaction had actually occurred was obtained when Enrico Fermi, directing the experiment, had control rods removed from the pile slowly over a period of hours. At each step of this process, Fermi carried out a series of calculations to make sure that the rate of neutron production remained under control. Finally, at 3:53 P.M., Fermi announced that a controlled nuclear reaction was taking place in the pile. He allowed the chain reaction to continue for just over four minutes before having the control rods dropped back into the pile, bringing the reaction to an end. Arthur Compton sent word of the experiment's success to his superior, James Bryant Conant, in a coded message saying that "The Italian navigator has just landed in the New World."[1] Scientists then knew that the only additional step needed to convert this pile into a weapon was to remove the control rods completely, allowing the nuclear chain reaction to go forward at a very rapid and uncontrolled rate. There was no thought of taking that next step with Chicago Pile One. Instead, the Met Lab received a new assignment: the design of a nuclear weapon based on plutonium, rather than uranium.

At that point in time, nuclear scientists had learned that only three isotopes undergo fission. As defined previously, isotopes are forms of an element that differ in their mass. For example, there are three isotopes of uranium, with atomic masses of 233, 235, and 238. They are represented by the symbols ^{233}U, ^{235}U, and ^{238}U, respectively. The three fissionable isotopes are ^{233}U, ^{235}U, and ^{239}Pu. The first of these isotopes exist in such small amounts in the Earth's surface as to be of no practical value in making nuclear weapons. The second isotope, ^{235}U, was the one used in constructing the Chicago Pile One. The third isotope, ^{239}Pu, is an isotope of an artificial element, discovered at the University of California at Berkeley in 1940.

An ongoing debate among scientists in the Manhattan Project focused on the question as to whether a uranium bomb or a plutonium bomb would be the better device for the first nuclear weapon to be built. In the end, a bomb of each type was built: Little Boy, the bomb dropped on Hiroshima, was a uranium weapon, while Fat Man, the bomb dropped on Nagasaki, was a plutonium bomb.

The use of these two bombs in August 1945 quickly brought an end to World War II. However, it marked only the beginning of efforts by the United States, the former Soviet Union (now Russia), China, and other nations to build more powerful and more efficient nuclear weapons. Over the next half century, an *arms race* developed in which these nations competed to catch up or stay ahead of each other in the construction of nuclear weapons. Today, seven nations are known to make up the so-called Nuclear Club. They are the United States, Russia, Great Britain, France, China, India, and Pakistan. These are nations that have developed and tested some type of nuclear weapon, either a fission bomb or a fusion bomb. (A fusion bomb is a nuclear weapon even more powerful than a fission bomb. It is made by combining, or "fusing," small atoms, rather than splitting large ones.) In addition to the members of the Nuclear Club, a number of other nations are thought to have built and/or tested some type of nuclear weapon. These nations include Israel, South Africa, and North Korea. In addition, as many as 40 other nations may have the capability of building nuclear weapons, although they have not yet done so.

Although the United States and other major powers have begun to find ways of bringing this nuclear race under some measure of control, that race has certainly not come to an end. In the early years of the 21st century, for example, both India and Pakistan were threatening to use their nuclear weapons in their decades-long dispute with each other. Also, the likelihood that North Korea has a well-developed nuclear weapons program has cast a pall over most of East Asia and has posed a difficult dilemma for the foreign policies of the United States, Japan, China, South Korea, and other nations in the area. When U.S. president George W. Bush announced the United States was to attack Iraq in early 2003, partially because it may have had a nuclear weapons program, the question began to arise as to how far the United States and other nations will go to prevent other nations, such as North Korea, from becoming new members of the Nuclear Club.

THE PEACETIME ATOM:
AN AGE OF PROMISE, 1945–1980

By the end of the 1940s, the threat of nuclear conflict was in the minds of nearly every U.S. citizen, if not of citizens throughout the world. But another aspect of nuclear power had also begun to rise in human consciousness: the potential peacetime applications of this new technology. Shortly after the end of World War II, the U.S. Atomic Energy Commission (AEC) published a booklet entitled *Atomic Energy: Double-Edged Sword of Science.* That booklet assured readers that nuclear energy could be used not only to make the most powerful weapons ever devised by humans but also to provide power for the generation of electricity; to propel ships, cars, trucks, and airplanes; and to make life better and easier in many other ways.

As the 1940s drew to a close, Americans and people throughout the world looked forward to a bright future that would be dramatically altered by the ready availability of inexpensive and environmentally safe nuclear power. Indeed, Lewis L. Strauss, then chairman of the Atomic Energy Commission, in a speech at the 20th anniversary of the National Association of Science Writers in New York City on September 16, 1954 had predicted that "Our children will enjoy in their homes electrical energy too cheap to meter."[2]

NUCLEAR ENERGY IN TRANSPORTATION

Strauss's forecast was supported by an event that occurred earlier that year. On January 21, 1954, the world's first nuclear submarine, the USS *Nautilus*, was launched at Groton, Connecticut. The man responsible for the development of the *Nautilus* was Admiral Hyman Rickover. At the end of World War II, Rickover had been posted to the Oak Ridge National Laboratory, in Oak Ridge, Tennessee, to learn more about nuclear technology. He quickly concluded that the technology could be applied to power submarines, and he began a campaign to convince government officials of his beliefs.

Over much of the decade, Rickover met significant opposition, both from naval personnel and from government officials responsible for oversight and funding of naval operations. Once the concerns of these officials had been satisfied, the design and construction of the *Nautilus* went forward rapidly. The *Nautilus* passed all its test missions with flying colors. One of the most appealing features of the nuclear-powered submarine was its ability to stay underwater for long periods of time. In theory, the *Nautilus* could have remained submerged for up to 11 years, although such an eventuality would never have occurred in fact. This and other attractive features of nuclear-powered submarines soon convinced naval experts that the era of conventionally powered submarines had come to an end.

The development of nuclear-powered surface ships, however, was less successful. The first such vessel, the NS *Savannah* (*NS* for "nuclear ship"), was launched in 1961. It served primarily as a demonstration of the way in which nuclear power could be applied to the construction of merchant ships. In the period between 1962 and 1970, it traveled around the world as a "goodwill ambassador" for the concept, sailing 430,000 miles (700,000 kilometers) without a refueling stop.

The *Savannah* proved not to be commercially feasible, however, costing significantly more to operate them comparable nonnuclear-powered vessels. It turned out to be both the first *and* the last nuclear-powered merchant ship built in the United States. (Germany and Japan each built a single nuclear-powered commercial vessel, but again they were both one-of-a-kind experiments.) The *Savannah* was decommissioned in 1981 and docked at Charleston, South Carolina, where it served as a museum until 1994. In that year, it was moved to Norfolk, Virginia, where it became part of the navy's reserve fleet.

Nuclear-powered military vessels became much more successful than the *Savannah* had ever been. Today, a number of the world's largest fighting ships are driven by nuclear reactors. Including its nuclear-powered submarines, the U.S. Navy has built and operated more than 80 nuclear-operated vessels, including the aircraft carriers *Enterprise, Nimitz,* and *Theodore Roosevelt;* the cruiser *Long Beach;* and the destroyers *Bainbridge* (decommissioned in 1996), *Truxton* (decommissioned in 1995), *Virginia* (decommissioned in 1994), and *California* (decommissioned in 1998).

The U.S. government and private industry promoted—and spent huge amounts of money in trying to develop—other military and peacetime applications of nuclear energy. Between 1946 and 1961, for example, considerable interest developed in the possible construction of a nuclear-powered airplane. Some engineers suggested that such an aircraft might be capable of flying more than 12,000 miles (20,000 km) at speeds of up to 450 miles an hour (700 km/hr) without, of course, having to stop for refueling. Enthusiasm for this project was based at least partially on the belief that the Soviet Union already had a well-advanced program for the development of nuclear-powered aircraft. As with so many other Cold War issues, the construction of such an airplane became yet another chip in the ongoing contest of superiority between the two great superpowers.

Thus, in 1946, the U.S. Congress authorized the creation of the Nuclear Energy Propulsion Aircraft (NEPA) project, later renamed the Aircraft Nuclear Propulsion (ANP) project. U.S. hopes for a nuclear-powered aircraft were eventually dashed, however, for two primary reasons. First, the nuclear reactor needed to drive such an airplane would have required lead shielding of significant size and weight. In fact, the amount of lead needed would have been so great as to make any nuclear-powered aircraft virtually unflyable. Second, engineers were unable to solve the problem posed by the possible crash of a nuclear-powered aircraft. In such a crash, the radioactive materials present in the reactor would be spread over a wide region, producing serious radiation hazards to humans and environments for hundreds or thousands of square miles. Fifteen years after the project had been inaugurated, therefore, Congress called an end to ANP. During its lifetime, the project had consumed more than $4.5 billion in tax funds, with no long-term useful results.

Projects aimed at the development of consumer products using nuclear power also experienced little success. In the 1960s and 1970s, for example, artificial heart pacemakers were designed and built using the radioactive isotope plutonium-238. An artificial pacemaker is an electronic device implanted into a person's chest that takes over the function of the body's natural cardiac pacemaker. An artificial pacemaker generates the electrical current it needs to operate from the radioactive decay of plutonium-238. The problem with the device is that plutonium is one of the most toxic substances known to humans. Although nuclear-powered pacemakers represented a relatively low risk to the individuals in whom they were implanted, their ultimate disposal posed a serious, long-term problem. Largely for this reason, and as more efficient chemical batteries became available, plutonium-powered pacemakers are no longer being produced in the United States.

In spite of its failure to meet expectations in some types of peacetime applications—such as merchant seagoing vessels, military and civilian aircraft, and a variety of specialized consumer products—nuclear energy has been far more successful in a number of other fields, including a variety of industrial and medical applications.

NUCLEAR ENERGY IN INDUSTRY

More than two dozen radioactive isotopes are currently in common use for a great variety of industrial applications. A radioactive isotope (or radioisotope) is an isotope that emits some form of radiation, such as alpha, beta, or gamma radiation, as it decays to produce a new kind of isotope. As an example, the radioactive isotope carbon-14 emits a beta ray as it decays to form the new isotope nitrogen-14.

These forms of radiation have different penetrating power. For example, an alpha particle is able to penetrate a thin sheet of paper, but its progress is stopped by a thin sheet of aluminum. A beta particle is able to penetrate the same sheet of aluminum but is stopped by a thin sheet of lead. A gamma ray is able to penetrate paper, aluminum, and thin sheets of lead but is stopped by a thicker piece of lead.

These properties mean that radioactive isotopes can be detected in systems even when they cannot be seen by the naked eye. When used in this way, radioactive isotopes are often known as tracers. The use of sodium-24 to detect leaks in pipelines is an example of the way tracers are used. Most pipelines are buried underground, making it difficult to know if and where a leak has occurred. If a small amount of radioactive sodium-24 is added to the fluid in a pipeline, however, leaks are easily discovered. As long as the pipeline is intact, no sodium-24 will escape to the surrounding soil, and no radiation will be detectable. If a leak occurs, however, some sodium-24 will escape into the soil, and the beta and gamma radiation it emits will be easily detectable with a Geiger counter or some other kind of radiation detection device.

A similar application of radioactive isotopes involves the process known as gauging, or determining the thickness of a material. Many industrial operations require that a sheet of material of relatively specific thickness be produced. Such operations include the manufacture of paper, plastic films, metallic plates, and rubber sheeting. For example, a manufacturer might wish to produce sheets of paper with a thickness of 2.50 mm ± 0.10 mm. A sheet of paper meets these standards if it is no thicker than 2.60 mm and no thinner than 2.40 mm. One way to ensure that this standard is being met is to stop the production process from time to time, measure the thickness of the paper, and then reset machinery if the paper is too thick or too thin. All the while, paper that does *not* meet this standard has to be destroyed or recycled.

The use of radioactive tracers provides a more efficient method for monitoring the thickness of a material. In this system, the material passes along a production line with a radioactive isotope above it and a radiation detection device beneath it. The detection device measures the amount of radiation that passes through the material as it rolls along the production line. If the material suddenly becomes thicker than the permissible standard, less radiation will get through to the detection device. If the material becomes thinner, more radiation will get through. In either case, the detection device can be calibrated to shut off the production line as soon as the irregularity in radiation is detected, preventing large amounts of the material from being wasted and allowing an adjustment in the production process immediately.

NUCLEAR ENERGY IN MEDICINE

Some of the most important applications of nuclear energy, in terms of volume of materials used and, even more important, benefit to human life, has been in the field of nuclear medicine. Nuclear medicine is that field in which radioactive isotopes and the radiation they produce is used for diagnostic and therapeutic purposes. Diagnosis is the process of searching for the cause (or causes) of a disease or disorder, while therapy is the process of ameliorating or curing the disease or disorder.

One of the first radioisotopes to be used in medical diagnosis was iodine-131. Iodine is preferentially absorbed in the body by the thyroid gland (and certain other organs). When iodine-131 is injected into the bloodstream, it tends to go almost entirely to the thyroid. By measuring the amount of radiation emitted by the iodine-131 absorbed by the thyroid, a diagnostician can determine whether the gland is functioning normally or not. Today, another radioisotope of iodine, iodine-123, is more commonly used in such studies.

Radioactive isotopes are used not only in the diagnosis of diseases and disorders but also in the treatment of those conditions. Recall that the radiation emitted by radioisotopes has the ability to ionize atoms and molecules, disrupting their normal functions. For this reason, such radiation is often called ionizing radiation. When ionizing radiation is directed at tumor cells, for example, it tends to ionize essential biochemical molecules in those cells, such as protein and nucleic acid molecules, destroying these molecules and killing the cells in which they exist.

Ionizing radiation, though, has the same effect on healthy cells that it has on tumor cells. Therapeutic use of ionizing radiation, therefore, always has some deleterious, as well as many beneficial, effects. One of the great challenges in the use of ionizing radiation for the treatment of disease, then, is to increase beneficial effects as much as possible while reducing the deleterious effects.

One method for achieving this goal is illustrated in the use of cobalt-60 in the treatment of cancerous tumors. Cobalt-60 is a radioactive isotope with a half life of about 5.27 years. It emits both beta and gamma rays, the latter being the primary agent for the destruction of tumors. A patient to be treated with cobalt-60 radiation lies on a table beneath a large donut-shaped machine. A cobalt-60 source is attached to a device capable of traveling in a circular path around the donut. Gamma radiation emitted by the cobalt source is directed by a computer at a specific location of the tumor in the patient's body. The gamma radiation always passes through some healthy tissue before it reaches the tumor, but it never passes through that tissue for long. By contrast, the radiation is directed so that it always passes through the tumor. In this way, the maximum amount of radiation reaches the cancerous area of the patient's body, and the minimum amount reaches healthy tissues in the body.

NUCLEAR POWER AS A SOURCE OF ELECTRICAL ENERGY

One can scarcely overestimate the important and, in some cases, revolutionary changes that nuclear energy has made in a variety of industrial operations and medical procedures. Still, the most significant long-term application of nuclear energy may well be in a totally different field: the generation of electricity.

Human civilization was transformed at the end of the 18th century with the invention of machines that operate by the combustion of fossil fuels, coal, oil, and natural gas. So profound was that change that the Industrial Revolution is sometimes said to have marked the beginning of the Age of Fossil Fuels. The modern world is absolutely dependent on an abundant supply of these fossil fuels. They are used to heat homes and office buildings; operate cars, trucks, trains, ships, and airplanes; power industrial operations; and drive an endless variety of other processes. Yet, the Earth's supply of fossil fuels is not infinite. At some time, humans will no longer be able to rely on coal, oil, and natural gas for their energy needs. To what new energy source will they be able to turn?

Almost from the moment the first fission bomb was dropped on Japan, a number of scientists, politicians, and industrialists began to consider the use of nuclear energy for the generation of power for peacetime applications. Accomplishing that goal would have any number of benefits, both economically and environmentally. A single kilogram of uranium generates about 3 million times as much energy as an equal mass of coal. And the generation of electricity from uranium produces none of the environmental problems of air and water pollution associated with the similar process for fossil-fueled electrical plants. At least in theory, then, nuclear power plants could provide energy to the public at less cost and greater safety to the environment.

NUCLEAR POWER TECHNOLOGY

A nuclear power plant is similar in design to a conventional power plant in which electricity is generated by the combustion of coal, oil, or natural gas. Diagrams of two common types of nuclear power plants are provided in Appendix F. The most important difference between the two types of power plants is the source of energy from which the electricity is generated: a fossil fuel in a conventional power plant and a nuclear reactor in a nuclear power plant.

The core of a nuclear power plant is, in fact, called just that: the reactor core. The reactor core contains a fissionable fuel, uranium or plutonium, from which energy is released. The fuel is fabricated in the form of long, thin cylindrical *fuel rods*, which are bundled together in groups called fuel assemblies. When uranium or plutonium atoms in the fuel fission, they release energy, which ultimately is used to boil water and produce the steam needed to generate electricity.

The reactor core is surrounded by a large dome-shaped building made of concrete and reinforced steel called a containment building. The purpose of the containment building is to prevent radioactive materials from escaping from the reactor core during operation and in case of an accident. Embedded in the reactor core along with the fuel rods are long, thin cylindrical control rods. Control rods are made of some element, such as boron or cadmium, that absorbs neutrons very efficiently. The position of the control rods with regard to the fuel rods determines the rate at which fission occurs with the reactor core. When the control rods are completely inserted into the core, large numbers of neutrons are absorbed, leaving too few to permit fission to continue. Were the controls rods to be removed completely from the core, very large numbers of neutrons would become available, and fission would occur very rapidly, in fact so rapidly as to create the danger of a serious accident in the power plant. In such an event, large amounts of heat energy would be generated, the reactor core would melt, and radioactive materials would be released. The reactor core would not, however, behave like a bomb and explode since the mass of fissionable material present is never large enough to permit this type of uncontrolled chain reaction. A core meltdown is, however, an extremely serious event. Clearly, the successful operation of a nuclear power plant depends absolutely on maintaining control rods in exactly the correct position to permit the release of energy from fission reactions at a safe and useable rate.

Once energy (in the form of heat) has been generated in the reactor core, some mechanism is needed to transfer it to an external site, where it can be used to boil water. Steam formed during the boiling of water is used to drive a turbine, which in turn runs the generator by which electrical energy is produced.

The two most common methods of transferring heat from the reactor core to the turbine are the boiling water system and the pressurized water system. In the former system, water circulating around the reactor core is allowed to come to the boiling point. The steam that is formed upon boiling is then transferred through pipes to the turbine. In the latter system, water circulating around the reactor core is kept under pressure, to prevent it from boiling. The super-heated water is then transferred through pipes to an external building where it is used to heat water in a second system, causing that water to boil.

Each type of nuclear power plant has its technical advantages and disadvantages. Today, about two-thirds of all nuclear power plants in the United States (69 reactors) use pressurized water systems, and the other third (35 reactors) use boiling water systems.

The superiority of one type of nuclear power plant design over the other (or, indeed, of any other type of nuclear reactor design) was not readily apparent to early researchers in the field of nuclear energy. Once the federal government had committed itself to the development of nuclear energy as a source of electrical power, therefore, a decision was made to construct a test site at which various reactor designs could be tested. It fell to the federal government to assume this responsibility since the cost and risk of carrying out such experiments were much greater than those that could be assumed by private industry.

The first research facility established in the United States was the National Reactor Testing Station (NRTS), located at Arco, Idaho (population: about 1,000), 40 miles east of Idaho Falls. Over succeeding years, the NRTS became the center of most basic and applied research on nuclear power production. Studies have been carried out there both by federal agencies, such as the Argonne National Laboratory, as well as by a host of private companies, including General Electric, Westinghouse, General Atomics, Aerojet General, and Combustion Engineering. The facility's name has undergone a number of changes since its founding in 1949 and is now known as the Idaho National Engineering and Environmental Laboratory (INEEL), representing a significant change in the laboratory's overall goals and activities.

The first project undertaken at NRTS was the design and testing of a breeder reactor. A breeder reactor, discussed later in this chapter, is one that generates more fuel than it consumes during its operation. At the time the project was initiated, many scientists and government officials had high hopes that breeder reactors might become an efficient source of nuclear power that would also be capable of generating new nuclear fuel, a hope that would, as it turned out, never be realized. NRTS also carried out research on nuclear reactors for use in submarines in response to Admiral Hyman Rickover's campaign to build underwater vessels powered by nuclear energy. Other nuclear projects that have been carried out at NRTS include research on reactor cooling systems, materials research for fusion reactors, development of radioactive isotopes, and studies of new and advanced types of nuclear reactors.

By the end of the 1990s, a total of 52 reactors had been constructed and operated at NRTS for studies on these subjects. Today, the amount of research on nuclear power reactors conducted at INEEL is greatly reduced, and studies on nuclear waste disposal problems has assumed a much more important role in the facility's activities.

One minor milestone was achieved at the station on December 20, 1951, when Experimental Breeder Reactor I (EBRI) produced the first electrical power from nuclear energy, providing enough electricity to light four light bulbs. Four years later, on July 17, 1955, Arco became the first town anywhere to obtain its electricity from nuclear energy, energy provided by an experimental boiling water reactor called BORAX III.

NRTS was also the site of one of the worst nuclear accidents in history, one of only two incidents in which human lives were lost in a nuclear facility in the United States. (The other accident in which lives were lost occurred at the Surry Nuclear Power Plant, near Norfolk, Virginia, on December 9, 1986.) On January 3, 1961, three technicians began routine maintenance operations of the reactor core of the experimental SL-1 (Stationary Low Power) reactor. The men accidentally raised the reactor's control rods too far, allowing the nuclear chain reaction to go out of control. The reactor almost immediately caught fire, releasing radioactive material to the room in which the reactor was located. Two of the men were killed immediately, and the third died shortly after the accident. The bodies of the three men were so radioactive that they had to be treated as if they were a form of nuclear waste and had to be buried in special sites designed to hold such materials. The reactor site itself was so radioactive that it could not be entirely decontaminated and repaired for 18 months.

BREEDER REACTORS

The seemingly illogical concept that a reactor can make more fuel than it actually consumes is based on the reaction that occurs between uranium-238 and neutrons in a reactor core. Recall that uranium-238 is not fissionable. Only the uranium-235 atoms in a fuel rod actually undergo fission. Uranium-238, however, *does* react with neutrons in an n,γ reaction to produce an isotope of the next heavier element in the periodic table, plutonium-239, which is one of the isotopes that *will* undergo fission. It can be removed from the "waste" products of either a boiling-water reactor or pressurized-water reactor, reprocessed, and used as fuel in a new nuclear reactor.

The possibility of building breeder reactors was an especially attractive idea to the United States and other nations in the 1950s because of the relatively limited supply of uranium-235. Uranium is not a particularly abundant element, and the extraction of uranium-235 (the fissionable isotope) from uranium-238 (the non-fissionable isotope) is a difficult, time-consuming, and expensive task. In the 1950s and 1960s, nuclear engineers hoped that they could find a faster, less expensive, and more reliable method of obtaining the fuel they needed from which to make nuclear weapons and operate nuclear power plants. Breeder reactors seemed to be the ideal solution to that challenge. By the early 1960s, the Atomic Energy Commission had invested about a half billion dollars in breeder reactor research.

For a time, engineers' hopes for breeder reactors appeared to be well-founded. The first commercial breeder reactor, the Fermi I plant at Lagoona Beach, Michigan, outside Detroit, began operation in 1966. The plant was in operation for only a few months, however, before an accident occurred. On October 5 of that year, for reasons that plant technicians did not then understand, the core temperature suddenly began to rise, forcing operators to shut down the plant's operations. It took nearly a year for engineers to discover the cause of the malfunction, a piece of metal that had come loose and blocked the cooling system.

Uncertainty about conditions leading to the metal fracture forced the plant's owner, Detroit Edison, to maintain the reactor on a stand-by status for nearly three years. Then, in 1969, the AEC gave Detroit Edison permission to restart the reactor. The plant operated normally for only a few months before another accident occurred in May 1970. The problem forcing the shutdown this time was quickly solved and the plant restarted two months later. By that time, however, Detroit Edison had begun to have second thoughts about the cost of operating Fermi I and the chances of further accidents. At the same time, the AEC had not yet issued permission for reopening the plant. As a result, Detroit Edison decided to close Fermi I permanently in 1972, leaving behind one of the most disastrous economic records in the history of nuclear power plants.

Fermi I's fate was not a sufficient warning for enthusiasts of breeder reactors in the United States and around the world. In the same year that Detroit Edison closed down its reactor (1972), the AEC announced an ambitious program to build two experimental breeder reactors. The first of these reactors was to be a liquid-metal fast breeder reactor to be constructed at Clinch River, Tennessee. The term *liquid metal* refers to the fact that such reactors were to use molten sodium metal as a coolant rather than the water used in most other types of reactors. The second project was designated as the Fast Flux Test Facility (FFTF), to be built at the AEC's research station at Hanford, Washington.

The Clinch River project was to be a cooperative program between the federal government and private industry, with the latter having pledged about one-third of the estimated half-billion-dollar cost of the experiment. Doubts about the project were expressed even before it was begun, with critics pointing out that its success depended upon unrealistically large increases in the demand for electrical power in the future and the virtual complete loss of uranium supplies.

Neither of these events occurred, and, in fact, the cost of uranium continued to drop, from a high of about $100 per kilogram in 1960 to about $50 per kilogram only a decade later. (The price has since leveled off at about $15 per pound [$33 per kilogram].) In addition, operators of the experimental plant encountered far more problems with the design and operation of the breeder reactor than they had anticipated. As a result, the project was canceled by Congress in 1983. At that point, $1.6 billion had already been invested in the project, about $240 million of which had come from industrial sources.

The Hanford project fared no better. Construction on the facility began in December 1970, and the reactor first went into operation in February 1980. When Congress decided to close down the Clinch River project in 1983, however, it also decided to change the focus of the FFTF facility, substituting the testing of fusion materials, the production of radioactive isotopes, testing for foreign governments, weapons research, and other projects above the development of a domestic breeder reactor. By 1990, Congress decided to discontinue funding for even these kinds of research, and two years later, dismantling of the FFTF began.

The U.S. experience has been repeated in a number of other nations around the world. France, Germany, Great Britain, Japan, Kazakhstan, and

Russia have all built large breeder reactors, operated them for a period of time ranging from three to more than 20 years, and eventually shut them down as having been unsuccessful methods of generating electrical power and producing plutonium. Today, there are no commercial breeder reactors operating in the world.

FUSION REACTORS

The fourth and most experimental type of nuclear reactor is the fusion reactor, a power plant built on the principle of controlled nuclear fusion reactions as the source of energy in the plant. Fusion reactions are nuclear reactions in which two small particles, such as two hydrogen atoms, are combined, or "fused," to form one larger particle. As with fission reactions, fusion reactions result in the release of large amounts of energy. Indeed, the amount of energy produced in fusion reactions is far greater than that obtained from fission reactions. Scientists now know that fusion reactions are the mechanisms by which stars produce energy.

Because of the enormous amounts of energy produced during fusion, a fundamental problem exists in the construction of fusion reactors that does not occur with fission reactors. Fusion reactors generate so much energy that they vaporize any form of matter known to humans with which they come into contact. The question becomes, then, how one might construct a fusion reactor that would not itself be destroyed by the energy produced within it.

One possible solution to this problem was suggested in the 1950s by two Soviet physicists, Igor Tamm (1895–1971) and Andrei Sakharov (1921–). The device they designed became known as a *tokamak*, the Russian word for "torroidal chamber." A toroid is a geometric figure with the shape of a doughnut. The key element in a tokamak is a very strong magnetic field that acts as a container within which a fusion reaction can occur. The problem of trapping the fusion reaction within a confined space is solved, then, not by using a form of matter, but a form of energy.

During the 1950s, the Soviets built a number of experimental tokamaks to test the design suggested by Tamm and Sakharov. Over time, interest in fusion reactors of this design had spread to other nations and, by the early 1990s, three large tokamak machines had been constructed outside the Soviet Union (and later Russia), each at a cost of more than a billion U.S. dollars each. They were the Tokamak Fusion Test Reactor (TFTR) at Princeton University, the Joint European Torus at Culham in the United Kingdom, and the Japan Tokamak 60 in Naka City, Japan. In addition, a number of smaller tokamaks had been constructed in Brazil, France, Germany, Japan, Russia, and the United States.

The primary challenge in commercializing fusion reactors has been to build a machine that generates more energy than it consumes. Fusion reactions involve the combination of like-charged particles, a process that requires enormous amounts of energy. In stars, such reactions occur only because temperatures are so high, a few million degrees, accounting for the name by which such reactions are sometimes known: thermonuclear reactions.

The first experiment in which a tokamak generated more energy than it consumed occurred at the Princeton facility on December 9, 1993. On that occasion, the TFTR produced a power output of 3 million watts for a period of one second. Nearly a year later, on November 2, 1994, the same reactor attained a maximum power output of 10.7 million watts, but again for no more than a few seconds.

In spite of the Princeton successes, the future of tokamak research at all three centers was not bright. It had taken three decades to realize the modest successes achieved at Princeton in 1993 and 1994, and government funding agencies began to raise serious questions as to how long they should continue supporting fusion research. In response to this crisis, fusion researchers in the United States, Russia, and other nations began to explore the possibility of creating a single large research center that could serve as the focus of fusion studies around the world. The product of that discussion was the creation of the International Thermonuclear Experimental Reactor (ITER) project, under the auspices of the United Nations Atomic Energy Agency. ITER consists of scientists and engineers from China, Europe, Japan, Korea, Russia, and the United States, working together to demonstrate the feasibility of fusion reactions as a source of commercial power. Long-term plans are for the construction of a single major test center (probably in Japan or Europe), aimed at the development of a machine capable of generating 1.5 billion watts of power for sustained periods of time.

THE POLITICS OF NUCLEAR ENERGY

The development of nuclear energy for peacetime applications such as nuclear power plants presented a host of scientific, technical, and economic issues. But it also created some difficult and fundamental political and legislative questions. In the late 1940s and early 1950s, it was not at all clear as to how the U.S. government should go about the process of promoting the development of nuclear-powered electrical generating plants. A number of difficult issues were involved in that process. In the first place, the technology needed for such programs, largely the same technology used for weapons development, was still classified information. It was not possible to hand out that information to just anyone who wanted to build a nuclear power plant to generate electricity. Also, private industry was reluctant, as it always is with a new technology, to spend large amounts of money on a technology about which so little was known, for which the future was uncertain, and which could result in huge financial losses.

The U.S. government's first attempt to solve this tangled issue was the Atomic Energy Act of 1946. The act was passed in response to a request by President Harry S. Truman issued less than two months after the end of World War II. The act is also know informally as the McMahon Act, after Senator Brien McMahon (D-Conn.), who sponsored the original legislation and shepherded it through an 11-month debate in Congress.

The long debate over the McMahon bill ranged over a number of topics related to nuclear energy, one of the most important of which was whether the new technology should fall under the control of the military or civilian agencies. On the one hand, General Leslie Groves, director of the Manhattan Project, and a number of other military officers argued that the threat of nuclear energy was so great that information about its technology should not be released to the general public or to the scientific community. That community, on the other hand, insisted that research had to continue on nuclear energy, not just on its military applications, but on the host of peacetime applications that it might also have.

Scientists were eventually able to convince McMahon of the legitimacy of their position, and his bill called for control of nuclear energy to be placed under civilian authority, in the form of an agency to be called the Atomic Energy Commission (AEC). The commission's primary task was to create a policy for domestic control of nuclear power. That policy was to be developed within a general mandate that all nuclear materials, facilities, and programs were to be controlled entirely by the U.S. government. In addition, the act prohibited the exchange of information about nuclear energy with any other nation.

An important provision of the 1946 act was the creation of the Joint Committee on Atomic Energy (JCAE). The committee was to consist of 18 members, nine from the Senate and nine from the House of Representatives. McMahon was chosen first chairman of the committee. The committee was unusual in one important way in that it had been established by statute, rather than by congressional rules, as is usually the case with congressional committees. JCAE had broad authority over virtually every aspect of the nation's nuclear energy policy, including both military and peacetime applications. The history of U.S. nuclear policy from 1946 to 1974 is, to a large extent, a story of the ways in which JCAE and the ACE worked together and, in some cases, in opposition to each other, to promote the development of nuclear energy in this nation and other parts of the world.

Nuclear Power

The 1946 act was passed over the objections of many corporations that wanted private industry to have greater access to nuclear information and a greater opportunity to develop private nuclear facilities. Passage of the bill did not bring that debate to an end, however, and, in fact, industry continued to lobby the U.S. Congress for a greater role in the peacetime development of nuclear power. Those efforts finally came to fruition in 1954 with the election of a Republican president, Dwight D. Eisenhower, and a Republican Congress, natural allies of business interests.

On February 17, 1954, President Eisenhower asked Congress to pass new legislation on nuclear energy, giving industry a larger role in its peacetime development. An act that incorporated these features was adopted by Congress on August 30 of that year, and subsequently signed by the president. It was named the Atomic Energy Act of 1954. The act ordered the AEC to provide private companies with the information they needed to build nuclear power plants, and it authorized companies to build, own, and operate such plants. The 1954 act essentially established the ground rules under which peacetime applications of nuclear power operates in the United States today.

One purpose of the Atomic Energy Act of 1954 was to encourage the construction of nuclear power plants by private industry. At the time, it was still unclear as to which nuclear power design—boiling water, pressurized water, or some other system—was the best choice for production of nuclear power. In an attempt to answer this question, the AEC in 1955 established the Power Demonstration Reactor Program (PDRP). The PDRP was a joint program between the federal government and private industry in which the AEC agreed to conduct basic research on nuclear power production in its national laboratories, to subsidize additional research carried out by private industry, and to provide the fissionable fuels needed by industry to operate nuclear power plants at no charge to companies. In return, private companies were to provide the financing needed for construction and operation of the demonstration facilities. The program was designed to last seven years.

The first contract signed under the PDRP was between the AEC and the Yankee Atomic Electric Company, based in Massachusetts. The contract was signed on June 4, 1956, and eventually resulted in the construction of the Yankee Nuclear Power Plant at Rowe, in western Massachusetts. Ultimately, seven prototype plants were constructed under the Power Demonstration Reactor Program before its shutdown in 1962.

Just prior to the adoption of the Atomic Energy Act of 1954, on December 8, 1953, President Eisenhower gave an address before the United Nations General Assembly, at which he announced that the United States was prepared to share with other nations of the world information that could be used for the peaceful development of nuclear power. In his address, Eisenhower pledged the United States's "determination to help solve the fearful atomic dilemma—to devote its entire heart and mind to find the way by which this miraculous inventiveness of man shall not be dedicated to his death, but consecrated to his life."[3]

This announcement dramatically revised U.S. policy on nuclear energy. Instead of continuing to reserve nuclear information exclusively for its own use, Eisenhower said, the United States was now willing to provide scientific and technical information and assistance to other nations of the world who wished to develop peaceful applications of nuclear energy. The president's offer was acclaimed throughout the world, but it had relatively little effect in the short term on actual practices of U.S. government agencies dealing with nuclear issues. Changes in those practices did eventually come about, however, and succeeding U.S. administrations have continued to be generous in their cooperation with other nations wishing to develop peaceful applications of nuclear energy.

The passage of the Atomic Energy Act of 1954 left only one major issue impeding the development of nuclear power: liability. In spite of aggressive public relations efforts by industry and government, almost everyone in the United States was aware that an accident at a nuclear power plant was possible. Mention of such a possibility always stirred memories of the detonation of Little Boy and Fat Man over Japan, events that occurred less than a decade earlier. According to a 1956 study by researchers at Brookhaven National Laboratory (the study is known as WASH-740 report), a nuclear power plant accident could result in about 3,000 deaths, 43,000 injuries, and property damage of about $7 billion. This report raised serious concerns among companies planning to build nuclear facilities.

To help resolve this issue, the U.S. Congress passed the Price-Anderson Act in 1957. The act was name after its two authors, Congressman Melvin Price (D-Ill.) and Senator Clinton Anderson (D-N.M.). The act limited a company's liability for any single nuclear accident to $500 million in government funds plus the $60 million that was available to a company from private insurance companies. The total liability a company would face in case of an accident was, therefore, set at $560 million. No matter what kind of accident a nuclear power facility might experience, then, victims could never collect an aggregate total of more than $560 million. The Price-Anderson Act was later extended and modified, most recently in 1988. In the 1988 amendment to the act, total liability was raised to $7 billion, an amount to be paid out of a fund maintained by fees paid by companies who own and operate nuclear facilities. The amendment relieved the U.S. government of any liability in the case of an accident.

By 1960, most of the roadblocks to the development of nuclear power facilities by private industry in the United States had been removed. As of

Nuclear Power

1960, only two nuclear power plants were in operation. The first plant to open was the Shippingport Nuclear Power Station, on the Ohio River, about 25 miles from Pittsburgh. The second plant was the Dresden Nuclear Power Station, in Morris, Illinois. The next two decades saw an explosion of plant construction in the United States. The number of nuclear power plants in the United States grew from two in 1960 to 12 in 1965 to 17 in 1970. Between 1970 and 1990, however, an additional 94 plants began operation.

During the 1970s and 1980s, a clear shift in the role of nuclear power in the United States's "energy equation" could be seen. In the early 1970s, only about 5 percent of the nation's electrical energy was generated by nuclear power plants. Over the next two decades, however, the fraction rose to about 11 percent in 1980, 15 percent in 1985, and 20 percent in 1990. A similar trend was observed in other parts of the world. Essentially an irrelevant factor in the generation of electrical power throughout the world in 1960, nuclear power grew to become one of the most important sources of electricity by 1990. Nuclear power plants now account for well over half the electrical energy generated in four nations—Belgium, France, Lithuania, and the Slovak Republic—and more than 40 percent in five other nations—Armenia, Bulgaria, Slovenia, Sweden, and the Ukraine. Additional data on the growth of nuclear power in the United States and other parts of the world are to be found in Appendix F.

The growth of nuclear power in the United States during the 1970s brought to the fore one final issue relating to the regulation of nuclear energy. At the time, control over all aspects of nuclear power was still a responsibility of the Atomic Energy Commission, created 25 years earlier. One ongoing criticism of the commission's work over that period of time, however, was the inherent conflict faced by the AEC: The agency was expected both to promote the use of nuclear energy and to regulate that use. It was as if the Food and Drug Administration were expected both to encourage the use of foods and drugs among the general public and, at the same time, issue regulations that restricted that use in some ways and to some extent.

This problem was finally resolved in 1974 with the passage of the Energy Reorganization Act (ERA). The ERA instituted a number of far-reaching changes in the way energy issues were managed in the U.S. government, one of which altered the way nuclear issues are handled within the government. The act created two new agencies, the Energy Research and Development Administration (ERDA) and the Nuclear Regulatory Commission (NRC). One of ERDA's many assignments was to promote the peaceful development of nuclear energy applications, while the primary responsibility of NRC was to monitor the planning, construction, and operation of nuclear power facilities, ensuring their safety and security. Only three years later, ERDA was abolished when the 1977 Department of Energy Organization Act created a new cabinet-level department by that name. Tasks originally assigned to ERDA were reassigned to the new Department of Energy (DOE). Today, most of ERDA's original mission has been assumed by the DOE's Office of Nuclear Energy, Science and Technology.

The decade of the 1970s seems, in retrospect, to have been the acme of nuclear power production in the United States. Americans had developed a new consciousness of environmental issues and of the nation's heavy reliance on fossil fuel energy imported from around the world. And, nuclear power plants then in existence appeared to be operating safely and efficiently. Hopes were high that nuclear power had begun to fulfill its promise of revolutionizing the way in which humans obtain the energy they need for their everyday activities, at least in the area of electrical production.

THE DREAM SHATTERS: NUCLEAR ACCIDENTS

The rosy picture of nuclear power soon proved to be an illusion. On March 28, 1979, an accident occurred at the Unit Two reactor of the Three Mile Island nuclear power plant in the middle of the Susquehanna River near Harrisburg, Pennsylvania. Although no lives were lost and no injuries attributable to the event occurred, the accident was to become the most serious setback to the development of nuclear power in the United States. Today, the story of nuclear power plants in this country can be divided fairly clearly into two distinct periods: pre–Three Mile Island and post–Three Mile Island events.

EARLY CONCERNS ABOUT NUCLEAR SAFETY ISSUES

The safety of nuclear reactors was a matter of concern to scientists and government officials almost from the moment that such facilities were first being seriously considered. Less than a year after it was formed, the Atomic Energy Commission established a group called the Reactor Safeguard Committee (RSC), whose function it was to advise the AEC on issues of reactor safety. The committee was chaired by physicist Edward Teller.

One of the RSC's first decisions was to take a somewhat different view of reactor accidents than was then popular among many experts, especially those who worked for the AEC. The prevailing view among these experts was that reactor accidents were highly unlikely and essentially not relevant to decisions about the construction, licensing, and operation of nuclear

29

power plants. Teller's committee decided to take the view that, however unlikely accidents might seem, some estimates should be made as to their possible consequences for human life and the surrounding environment.

In its first reports, the RSC made two major recommendations. First, it suggested that nuclear power plants should be built as far from heavily populated areas as reasonably possible. Second, it recommended that research be initiated to provide sound scientific data about the probability of an accident's occurring and the effects that might be expected from a reactor accident. The creation of the NRTS in Idaho in 1949 was a result, at least in part, of the committee's recommendations.

The RSC was reconstituted in 1953 and renamed the Advisory Committee on Reactor Safeguards (ACRS). Since that time, the ACRS has functioned as a watchdog on nuclear safety issues, first for the ACE, and later for its successor, the Nuclear Regulatory Commission. The relationships between the RSC (and ACRS) and its parent groups has not always been congenial. In the early 1960s, for example, the ACRS strongly opposed licensing of the Fermi I nuclear power plant in Lagoona Beach, Michigan. The committee expressed concerns about the possibility of a major accident at the site, citing the unknown and potentially hazardous nature of the breeder reactor to be used at Fermi I. The AEC overruled its advisory group and allowed construction of the plant. Although no major accident occurred at the plant, it experienced a series of serious accidents and disruptions and operated for only short periods of time before its closing in 1972.

Even as the Reactor Safeguard Committee was beginning its work, the AEC was receiving the earliest reports about possible consequences of a nuclear power plant accident. The first of these reports was issued in March 1957 by staff scientists at the Brookhaven National Laboratory, on Long Island, New York. The report, generally known as the WASH-740 report (its name simply representing a report number) estimated the damage that might occur as the result of an accident at a nuclear power plant of average size (165 megawatts) located 30 miles upwind of a typical American metropolitan area. Authors of the report predicted that 3,000 people would be killed, 43,000 more would be injured, and property damage would amount to approximately $7 billion.

The report was a disaster for the AEC. At a time when the government was trying to interest private industry in constructing nuclear power plants, the Brookhaven estimates cast a chill over the prospect of future involvement by private businesses. The AEC continued to argue that the chances of a nuclear power plant accident were vanishingly low, but industry was not convinced. In fact, it was not until passage of the Price-Anderson Act in 1957 that private companies once again began seriously to consider a role in the development of nuclear power plants.

Meanwhile, AEC officials continued to agonize over the Brookhaven report and the image it gave of nuclear power plant safety. The commission decided to request a second study on the question, hoping that it might provide a more positive view than the one expressed in WASH-740. Again, the research was assigned to AEC scientists at Brookhaven, who, in 1965, issued their report, the WASH-740 update. In this report, researchers made two new assumptions, both based on recent trends in the nuclear industry. Those assumptions were that (1) nuclear power plants would be larger than those envisioned in the original report (1,000 megawatts rather than 168 megawatts), and (2) those plants would be located in closer proximity to metropolitan areas than had originally been assumed. With these new assumptions, Brookhaven researchers predicted that a reactor "accident" would result in 45,000 deaths, 100,000 injuries, and $17 billion in property damage.

When AEC officials received this report, they questioned its scientific validity and sought to have its results reassessed. In its paradoxical role of promoter and regulator of nuclear power, the AEC chose to come down on the former side of its responsibilities. It decided to make every possible effort to encourage the construction of nuclear power plants, even given the potential dangers posed by such facilities. When the fundamental conclusions of the WASH-740 update report were eventually confirmed, the AEC decided not to publish the report, concluding that its effect on public opinion might be disastrous for the nuclear industry. The findings of the WASH-740 update report are still not available to the general public.

A fundamental problem with both the original WASH-740 report and its update was that their authors had not been able to quantify the probability of a nuclear accident. The possibility of 3,400 or 45,000 deaths could not properly be assessed without knowing the likelihood of a nuclear reactor accident. Those numbers had meaning only if there was reason to believe the chance of an accident was 1 in 10, 1 in a million, or 1 in 10 billion.

To remedy this shortcoming, the AEC commissioned a new study in 1974 to be conducted by Norman Rasmussen, then professor of nuclear engineering at the Massachusetts Institute of Technology (MIT). Rasmussen developed a complex and sophisticated mathematical system that he called the Probabilistic Risk Assessment (PRA) method for estimating the probability of an accident at a nuclear power plant. The system involved estimating the ways in which various combinations of events might lead to an accident, the probability of each event in a combination's occurring, and the overall probability of the set of event's taking place. The result of this study, which came to be known as the Rasmussen Report (also known as WASH-1400), was issued on October 30, 1975. It concluded that the probability of a nuclear accident was very small, less (often much less) than other common risks, such as airplane crashes, earthquakes, tornadoes, hurricanes, dam failures, major fires, and explosions.

The Nuclear Regulatory Commission (which had assumed AEC's regulatory duties by that time) was overjoyed with Rasmussen's conclusions. The report seemed to justify the commission's long-held contention that nuclear power plants were, for all practical purposes, entirely safe, and that the general public had no cause to worry about them as a reliable source of electrical energy.

Other organizations were less enthusiastic about the Rasmussen Report. These organizations raised questions about Rasmussen's basic assumptions, methodology, mathematical calculations, and conclusions. Perhaps most damaging of all was the report of a committee of the American Physical Society (APS) appointed to assess the Rasmussen Report. The APS committee pointed out that Rasmussen had, among other things, neglected the long-term health effects of radioactive contaminants, such as cesium-135 and cesium-137. That isotope alone would, APS reviewers pointed out, be responsible for an unknown—but undoubtedly large—number of fatalities from cancer in years following a nuclear accident.

In January 1979, the NRC issued a statement acknowledging the legitimacy of many of the criticisms aimed at the Rasmussen Report. To a significant degree, the commission found itself essentially where it had been prior to the Rasmussen study, with little sound understanding of the likelihood that a nuclear power plant accident might occur.

The way in which the WASH-740 update and Rasmussen reports were handled by the AEC and NRC contributed to a growing feeling among many observers that neither agency had chosen to deal forthrightly with issues of nuclear reactor safety. Neither the AEC nor the NRC appeared to have sufficient information about the health and environmental effects of radioactivity, the willingness to share the information it did have, the commitment to studying safety issues, or the resolution to place health and environmental concerns above those of industry profit. It was this mistrust of the AEC and NRC that was to play an increasingly important role in the development of a large, vocal, and successful antinuclear movement in the 1970s. All that was needed to unleash the floodgates of that movement was a single event that made the risks of nuclear power a reality in the lives of ordinary citizens.

NUCLEAR POWER PLANT ACCIDENTS

The concerns expressed about the safety of nuclear power plants expressed in the two WASH-740 reports seemed, for 15 years, to be a matter of theoretical concern, of little relevance to the day-to-day issues of nuclear power plant construction and operation. That situation changed suddenly and dramatically on March 28, 1979, when news of a serious accident at the Three Mile Island nuclear power plant was made public. The Three Mile Island (TMI) accident, as is the case with many industrial accidents, resulted from a combination of human and mechanical errors. During a routine maintenance operation, a safety device controlling the flow of cooling water through the reactor core failed, shutting down the circulation of water. Temperatures in the reactor core rose quickly, creating the possibility of a meltdown.

To some people, a meltdown seems comparable to the uncontrolled fission reactions that take place in a nuclear weapon. They imagined that a nuclear power plant that experiences a meltdown will explode like Little Boy or Fat Man. But such is not the case. The reactor core in a nuclear power plant never contains enough uranium or plutonium to permit the kind of explosion that occurs with a nuclear weapon.

A meltdown is dangerous enough in its own way, however. For example, large amounts of radioactive gases are released from the reactor core. And, in theory, enough heat is generated to blow the core and the surrounding building apart, releasing those gases to the atmosphere. To prevent such an event, the core and its related components are housed within a containment dome, a steel-reinforced concrete structure designed to withstand almost any explosion that might occur within the core.

At Three Mile Island, the reactor core became very hot, although it did not undergo a complete meltdown. Enough heat was generated, however, to cause radioactive gases to be released from the reactor core. Those gases then escaped from the containment dome through the plant's normal ventilation system, an event of which plant managers were not aware for a period of more than 12 hours. During that time, these dangerously radioactive gases were released into the atmosphere above and around the plant.

Even after plant managers had discovered this release of gases, it took them another 12 hours before the gases were contained. At that point, they thought that the accident had been brought under control. Such, however, was not the case. Managers next discovered that a huge bubble of hydrogen gas with a volume of about 30,000 liters (1,000 cubic feet) had collected inside the containment building above the reactor core. Ignition of the gas would have resulted in a huge explosion, causing damage of unknown—but serious—consequences. As a result, managers recommended evacuation of an area surrounding the plant extending about 15 kilometers (10 miles) in every direction. Pennsylvania governor Richard Thornburgh decided to ignore that recommendation and, instead, ordered the evacuation of pregnant women and preschool children within a two-kilometer (one-mile) radius of the plant. He appeared to have had less concern about the risk to other individuals living within the area and to everyone outside the two-kilometer perimeter. Fortunately, Thornburgh's optimism was justified when the hydrogen bubble gradually disappeared over the following few days.

The Three Mile Island accident was the worst nuclear power plant accident in U.S. history. But it was neither the first nor the last of such events to occur in this country and other parts of the world. In November 1955, for example, an experimental reactor at the National Reactor Testing Station (NRTS) in Idaho Falls, Idaho, experienced a partial meltdown. Six years later, a similar accident occurred in a second reactor at NRTS, releasing radiation into the surrounding area. In October 1957, a reactor at the Windscale plant north of Liverpool, England, caught fire, resulting in the release of radiation to the surrounding environment. Similar events occurred at the Chalk River Nuclear Power Station in Ontario, Canada, in 1958; at the Enrico Fermi plant in Detroit in 1966; in Saint-Laurent, France, in 1969; in Shevchenko in the then–Soviet Union in 1973; in Decatur, Alabama, in 1975; and at the Rancho Secco plant near Sacramento, California, in 1978.

By far the most serious nuclear accident in history, however, was the one that took place in Unit 4 of the Chernobyl Nuclear Power Plant, near Kiev, in the Ukraine. During routine safety tests on the reactor core of Unit 4 on April 25, 1986, two different maintenance crews made a series of mistakes that caused the reactor core to overheat and, eventually, to explode. (The explosion was a chemical explosion, not a nuclear explosion.) Later studies indicated that about a quarter of the core, with a mass of about 500 metric tons, was expelled into the atmosphere. The highest temperature reached during the explosion was also later estimated at no less than 2,225°C.

Fire quickly spread throughout the plant and was extinguished only through the heroic efforts of 186 firefighters from 37 different stations. The release of sand, clay, and dolomite on the burning plant by Soviet bombers also contributed to extinguishing the fires. Unfortunately, one of the many tragic consequences of the Chernobyl disaster was the loss of life among firefighters and other rescue workers who were exposed to very high levels of radiation in the days following the fire. In less than a week following the explosion, 29 firefighters lost their lives to radiation sickness.

Control of the fire did not, however, end the threat posed by the explosion. Radioactive gases released from the reactor core were blown to the northwest by prevailing winds, carrying them as far away as Great Britain. Officials later estimated that the total amount of radiation released in the explosion was 200 times that produced by the two bomb explosions over Hiroshima and Nagasaki in 1945. The number of deaths attributable to radioactive poisoning is virtually impossible to know, although the Ukrainian government has estimated that number at 4,229 for the period between 1986 and 1996. Some authorities question the reliability of that figure and suggest that the actual number of deaths from radiation poisoning was much greater.

The number of individuals killed as a result of the Chernobyl accident, however large it may be, hardly begins to estimate the overall damage caused by the accident. By some estimates, as many as 10 million people in Europe may have experienced at least some level of health problems as a result of exposure to radiation. In the regions closest to Kiev, such health problems have been studied in considerable detail and found to include an increased rate of thyroid cancer among children and an elevated number of birth defects among newborn children in Ukraine and the neighboring country of Belarus. Fallout from the explosion has also seriously contaminated large areas of the ground in the Ukraine, Belarus, and Russia. According to some estimates, 21 percent of the ground in Belarus is contaminated with one of the most dangerous radioactive isotopes, cesium-137, and Belarusian officials estimate that 16 percent of their land will still be contaminated in 2016.

The Ukrainian government's "final solution" to the Chernobyl accident was to construct an enormous concrete sarcophagus that would completely cover the Unit 4 plant. The sarcophagus was designed to last for hundreds of years, trapping (it was hoped) the 740,000 cubic meters (26 million cubic feet) of radioactive material, including 97 percent of the plant's original fuel rods. In less than five years, the failure of that plan became apparent. With temperatures inside the sarcophagus still at levels of more than 200°C, the concrete shell and its supporting pillars began to weaken and break apart. By 2004, dozens of holes and cracks, covering more than 1,000 square meters (10,000 square feet) had opened up in the face of the sarcophagus. Concern began to grow among the international community about what some experts called "one of the most dangerous nuclear facilities in the world."[4]

In April 1996, a conference was held in Vienna, Austria, at which nuclear experts from around the world initiated a discussion as to the steps that could be taken to make the Chernobyl site more safe. Out of that conference came a plan in which the French and German governments agreed to work with officials of the Ukraine to construct a new, larger, and stronger shelter for the Unit 4 reactor. That shelter would consist of a 20,000 ton steel box, 100 meters by 120 meters by 260 meters (about 300 feet by 350 feet by 800 feet) in size, about as tall as a 37-story office building. The box is to be built on open land adjacent to the sarcophagus and then slowly slid into place around the deteriorating concrete structure. Plans call for completion of the project in 2008.

GROWTH OF AN ANTINUCLEAR MOVEMENT

Opposition to the use of nuclear energy for both military and peacetime applications has existed from the earliest days of the atomic age. At the conclusion of World War II, that opposition was directed against nuclear weapons testing and development. During the late 1940s, the memories of the horrible attacks on Hiroshima and Nagasaki were still fresh in the minds of men, women, and children around the world. And the terrible risks of atmospheric testing of weapons that was part of the arms race between the Soviet Union and the United States soon became all too apparent. A large and vigorous movement opposing the development and use of nuclear weapons grew up in nations around the world. That movement had only modest impact on the arms race, however, with political factors playing a more important role in the continued development of nuclear arsenals in the United States, the Soviet Union, Great Britain, France, and other nations. Many of the antinuclear weapons groups, however, later evolved into or developed subsidiary programs in opposition to nuclear power plants. For example, the Greater St. Louis Committee for Nuclear Information was created in 1958 to provide information about the use of nuclear energy in weapons development and use. Eventually, however, the organization also embraced questions of nuclear power plant safety and operation.

Early proposals for the development of peacetime applications of nuclear energy met with mixed responses. It was virtually impossible, to some extent, to separate the concept of "nuclear energy" as used for power plants with the kind of nuclear bombs dropped on Japan in 1945. The fears of a nuclear holocaust engendered by the events of Hiroshima and Nagasaki transferred naturally in many peoples' minds to the risks posed by a nuclear power plant.

On the other hand, the potential benefits of nuclear energy for the modern world were difficult to ignore. For those concerned about the environment, for example, nuclear power plants appeared to represent the best available technology for replacing the smoke-belching, pollution-generating, health-endangering fossil-fueled power plants that environmentalists so strongly opposed. During the 1950s and early 1960s, then, most environmental groups endorsed, if somewhat uneasily and often with reservations, the construction of nuclear power plants. The Sierra Club, for example, worked with Pacific Gas and Electric (PG&E) during the early 1960s in an effort to find a suitable location for a nuclear power plant that the utility intended to build. At the time, David Brower, executive director of the club, expressed the view that nuclear power was "an environmentally benign alternative to dams and traditional power plants."[5] The club eventually signed off on PG&E's selection of Diablo Canyon for its plant, although the site was known to contain a major earthquake fault.

Many ordinary citizens and public interest groups also supported the development of nuclear power, having been convinced by claims like those of Strauss (cited earlier in this chapter) that nuclear-generated electricity would be almost free and entirely safe. As late as 1981, for example, the Methodist Church of England was still encouraging the development of nuclear power in its report "Shaping Tomorrow." That report claimed that

1. Nuclear energy is an integral part of nature, just as much God's creation as sunshine and rain.
2. It does offer mankind a new energy source which is very large, convenient and not very costly.
3. Around the world, the most important energy sources, oil in the rich world and wood in the poor, are becoming scarce, so that we cannot afford to set aside any technology with large potential which is cost effective, provided it is reasonably safe.
4. There are risks associated with the use of nuclear power, as with everything else, but these have been very carefully evaluated, are not very big and are not at all out of scale compared with risks of other energy sources and other ordinary hazards.[6]

Overall, support for nuclear power plants in the United States tended to win out over opposition by relatively large margins in polls taken in the 1950s, 1960s, and early 1970s. Some of the best long-term data available come from Cambridge Reports, Inc., which has been tracking public opinion about nuclear power plants since the early 1970s. Polls conducted by the company during the early years of this period showed that Americans favored the construction of nuclear power plants by a majority of about two to one, with a significant number of respondents (about 20 percent) having not yet made up their minds about the question. Until 1979, according to one analysis of these data, "opposition levels [to nuclear power plants] averaged 25 to 30 percent, indicating that substantial majorities of the public favored further nuclear development."[7]

Organized opposition to the development of nuclear power in the period between 1950 and 1979 was largely local and sporadic. At various locations across the country, small groups of citizens organized to oppose the construction of specific nuclear power plants because of their potential aesthetic, environmental, and/or health effects.

One such example was the successful effort by citizens of northern Indiana to prevent the construction of a nuclear power plant on the shores of

Lake Michigan near the town of Bailly, Indiana. The Northern Indiana Public Service Company (NIPSCO) had proposed in 1972 the construction of Bailly Nuclear I in order to meet projected increases in demand for electrical energy over the coming decades and to protect its customers against the increasing cost of electricity produced by fossil-fueled plants resulting from higher oil prices. NIPSCO had chosen to build the plant adjacent to the Indiana Dunes National Lakeshore in the northwestern corner of the state.

Residents of the Dunes area were less than enthusiastic about NIPSCO's plans. Some were concerned about possible releases of radiation from the plant. Others worried about the contamination of lake water by wastes from the plant. Still others objected to disruption of the national beauty of the Dunes area by the plant and its operation. A number of organizations were formed to oppose the construction of Bailly I Nuclear, the largest of which were the Save the Dunes Council and the Bailly Alliance. These organizations made use of the protest tools available to such groups, including public meetings, forums at local colleges, speeches by outside experts, newsletters and other publications on general and specific nuclear-related issues, sales of T-shirts and buttons, and a well-attended two-day Midwest No-Nukes Conference. In addition, the antinuclear groups undertook a series of legal maneuvers, directed first at the NRC and later the federal court system, in an effort to preventing NIPSCO from proceeding with its plans.

In one respect, the anti-NIPSCO groups lost their battle when the U.S. Supreme Court ruled on November 11, 1975, against their objections to the Bailly plant in *Northern Indiana Public Service Co. v. Porter County Chapter of the Izaak Walton League of America, Inc., et al.* (423 US 12). In a larger sense, however, they won when NIPSCO decided that the Supreme Court's decision was only a single victory in what would probably be a much larger war over Bailly Nuclear I. As a result, the utility announced on August 26, 1981, that it was abandoning plans for construction of the plant. After a decade of dispute and legal maneuvering, all that remained of the project was a large hole in the ground on the shores of Lake Michigan dug at a cost to the utility of more than $200 million.

Perhaps the best known and most widely studied nuclear power plant protest of the late 1970s was that which developed over the proposed construction of a reactor by the Public Service Company of New Hampshire (PSNH) in the town of Seabrook, about five miles north of the Massachusetts–New Hampshire border. PSNH announced its intention to construct a nuclear power plant in December 1968. Those plans met with modest objections from a local environmental group, the Sea Coast Anti-Pollution League, objections that became moot when PSNH decided to build a fossil-fueled plant rather than a nuclear facility. The nuclear option did not die, however, and nearly a decade later, PSNH once again announced plans to construct a nuclear power plant at Seabrook. This time, objections to the utility's plans were much more vigorous, especially when groundbreaking actually began for the facility. Protestors created an organization that they called the Clamshell Alliance, a name that was to go down in the history of antinuclear protests in the United States and throughout the world.

The Clamshell Alliance mounted its first protest meeting on the site of the proposed Seabrook plant on May 1, 1977. More than 2,500 men, women, and children attended that protest, designed to be a nonviolent act of civil disobedience against construction of the plant. When protestors refused to leave the site, police arrested 1,414 of the demonstrators, all of whom were jailed for periods of up to 12 days awaiting trial. This apparent defeat for the Clams (as they came to be called) only strengthened their resolve to fight PSNH's plans for the Seabrook nuclear facility. They continued to conduct protest meetings at the site of the plant over the next decade. In some cases, those protests became violent. In 1979, for example, police used tear gas, attack dogs, and riot equipment to remove more than 2,000 protestors from the construction site. The largest rally mounted by the Clams was held at Seabrook on June 8, 1978, when more than 10,000 people showed up to demonstrate against the proposed facility.

Actions of the Clams delayed, but did not stop, completion of the Seabrook plant. Even when construction had ended, however, the PSNH's problems were not over. Governor Michael Dukakis, of neighboring Massachusetts, refused to cooperate in a program of emergency planning exercises required before the NRC would license the plant. Finally, in 1990, the NRC decided to issue an operating license without the cooperation of Massachusetts officials, arguing that the state would certainly cooperate in case a real emergency were ever to occur. Seabrook continues to operate today under the management of FPL Energy.

The Bailly Alliance and Clamshell Alliance were representative of the way in which individuals organized locally during the 1970s to work against the construction of nuclear power plants in or near their own communities. Similar groups were formed in California (the Abalone Alliance), 10 southern states (the Catfish Alliance), South Carolina (the Palmetto Alliance), Louisiana (the Oystershell Alliance), Washington (the Crabshell Alliance), and Kansas (the Sunflower Alliance). These groups met with varying degrees of success, usually in delaying and greatly increasing the costs of construction, but seldom preventing completion of a project. Once their specific case had been won or lost, the groups usually disbanded and disappeared from the antinuclear movement.

Even as groups like these were organizing opposition to nuclear power on a local level, hints began to appear of broader mechanisms by which such opposition could be created. In 1974, for example, consumer advocates Ralph Nader and Joan Claybrook, founders of the public interest group Public Citizen, convened the nation's first antinuclear power conference, Critical Mass '74. More than 1,200 antinuclear activists from around the country attended the meeting, where they were instructed in the basics of carrying out grassroots campaigns against nuclear power projects. The success of the conference prompted Nader and Claybrook to form a special interest group within Public Citizen, a group also called Critical Mass, to coordinate and support opposition against nuclear power. They also organized a follow-up conference a year later, Critical Mass '75. Critical Mass continues to function today within Public Citizen with the goal of "protect[ing] citizens and the environment from the dangers posed by nuclear power and seek[ing] policies that will lead to safe, affordable and environmentally sustainable energy."[8]

The antinuclear movement also received a somewhat unexpected boost in 1972 with the revelations of potentially serious generic problems with nuclear power plant design. These problems involved a backup set of components and procedures known as the Emergency Cooling Control System (ECCS). A plant's ECCS is designed to provide a last-resort protection against meltdown in case the facility's primary cooling system fails. In response to safety standards announced by the AEC in 1971, a number of scientists at government laboratories raised questions as to whether existing ECCS designs were adequate to protect against major accidents. The AEC held hearings on this question and found experts in the field in substantial disagreement. Some felt that the ECCS used at existing nuclear power plants was entirely adequate, while others saw serious flaws in the design that had been adopted and was then accepted as the standard for all nuclear reactors.

The problem for the nuclear industry was that opponents of nuclear power saw this debate as further justification for their concerns about the safety of nuclear reactors. In addition, those opposed to nuclear power plants were convinced that the AEC had intentionally covered up information they felt would be harmful to the industry. The AEC was placing the best interests of the nuclear industry, they said, over the health and safety of the general public. As Alvin Weinberg, then director of the Oak Ridge National Laboratory, said at the time, "It makes me all the more unhappy that certain quarters in the AEC have refused to take it (ECCS problems) seriously until forced by intervenors who are often intent on destroying nuclear energy!"[9]

The significance of the ECCS debate for the nuclear power industry was that it not only revealed a significant design problem with nuclear power plant technology and the AEC's duplicitous approach to dealing with the problem, but also, probably for the first time, raised serious questions about nuclear power production that went well beyond local issues and that involved national policy about nuclear power.

The mid-1970s also saw the growth of regional and statewide campaigns against nuclear power. In most cases, these campaigns took the form of initiative proposals placed on state ballots by opponents of nuclear power. These proposals called for a halt on new construction, cutbacks on or closing of operations of such plants, improved safety systems, adoption of waste disposal plans, or some combination of these features. In 1976 alone, seven states—Arizona, California, Colorado, Montana, Ohio, Oregon, and Washington—voted on such initiatives. The proposals were defeated in all seven states by margins ranging from 71 percent to 29 percent (in Colorado) to 58 percent to 42 percent (in Montana and Oregon). Only in California, where the state legislature placed certain conditions on the construction of nuclear power plants in 1976, did such statewide measures achieve any level of success.

Indeed, a disinterested observer might be hard put to assign much credit to the efforts of antinuclear power activists prior to the late 1970s. The construction of some plants was delayed, some plants were actually canceled, and all construction was made much more expensive. But technical problems encountered by the nuclear industry itself (such as those that marked the death knell of Fermi I) were probably at least as responsible for delays and disruptions in plant construction as were the actions of opponents to nuclear power.

In spite of opposition from antinuclear groups and its own internal problems, the nuclear industry experienced its greatest success in history in the decade from 1965 to 1975. During that period, 224 nuclear power plants were ordered by industry. In the period of 1972 to 1974 alone, 108 new orders were placed, more than have been ordered in all of history before and since. Nuclear energy seemed on its way to a growing and significant role in the U.S. energy equation as the 1970s grew to a close. Then came the Three Mile Island accident of March 28, 1979, and the antinuclear movement was suddenly and spectacularly revitalized.

EFFECTS OF THE TMI ACCIDENT

One of the first concrete effects of TMI was a reversal in public attitudes about nuclear power. Support for the technology dropped from a high of about 60 percent in 1977 to a low of about 30 percent only five years later. During the same period, opposition to nuclear power generation rose from about 30 percent to nearly 60 percent. The impact of TMI became apparent within months of the accident. A march in Washington, D.C., opposing nuclear power on May 7, 1979, drew an estimated 65,000 people, with contingents from a wide variety of groups ranging from the Communist and Socialist Workers Parties to the Union of Concerned Scientists to the Grey Panthers to the Gay Liberation Movement. Later the same year, a rally at Battery Park in New York City drew an estimated 300,000 people.

Perhaps most significantly, opposition to nuclear power had spread to corporations and utilities. In 1980, for example, a group of General Electric (GE) stockholders formed GE Stockholders Alliance against Nuclear Power, a group working to convince the giant company to withdraw from its huge nuclear power and nuclear weapons programs. At about the same time, a consortium of 30 municipal utilities in western Massachusetts withdrew more than half its financial commitment to the Seabrook nuclear project.

Some regulators also added to nuclear industry woes by shifting financial responsibility for accidents from customers to shareholders. A 1979 ruling to that effect by the Vermont Public Service Board prompted two of the state's utilities to withdraw from the Seabrook project, contributing even more problems to the owner's efforts to get that facility into operation. Opponents of nuclear power were also once again emboldened by TMI to take their case to the public in state initiatives and referenda. Proposals similar to those that had been submitted in seven states in 1976 (and that had all failed) were placed on the ballots in Montana in 1978; Maine, Missouri, and Oregon in 1980; Washington in 1981; and Idaho, Maine, and Massachusetts in 1982. This time around, opponents were more successful, winning in Montana, Oregon, Washington, Idaho, and Massachusetts.

Probably most important of all, the NRC and other regulatory agencies began to add layer upon layer of licensing rules, construction provisions, and operation regulations that greatly increased the cost of building and running a nuclear power plant. Costs of completing plants under consideration or under construction mushroomed to three, four, or more times their original projected costs. Most power companies began to question whether such plants were economically viable. They began to abandon plans for the construction of nuclear facilities in favor of more traditional, less-expensive fossil-fueled power plants. For example, one of the last nuclear power plants to be licensed, the Midland (Michigan) Nuclear Reactor owned by Consumers Energy, was converted to natural gas operation in 1990 when its owners decided that regulatory costs and public opposition to nuclear power were too great to make it economically viable as a nuclear facility. The irony was that the well-documented health risks of fossil-fueled plants were many times greater than the health risks that had been posed by nuclear powered facilities in the previous decades.

The return in the United States of dependence on fossil fuels, rather than nuclear energy, occurred at a time, perhaps not entirely coincidentally, when the nation's energy policy was under the control of presidential administrations (Ronald Reagan, George H.W. Bush, and George W. Bush) with close ties to the fossil-fuel industry. The latter two presidents had, in fact, spent much of their adult lives, and had made their personal fortunes, in the oil industry. They had also appointed to their administrations others with similar backgrounds. The single intervening Democratic administration, that of Bill Clinton, had done little to change the course of the nation's rejection of nuclear power and dependence on fossil fuels.

As a consequence of this shift in attitudes and national policy, the role of nuclear-generated electrical power in the United States began to level off in the mid-1980s. The percent of electricity generated by nuclear power rose from 11 percent in 1980 to nearly 20 percent by 1988. It then leveled off and remained nearly constant over the next 15 years. The number of nuclear power plants in operation in the United States has also remained essentially constant for the last 20 years at slightly more than 100. The last year in which a new nuclear power plant was ordered was 1978, and no new plant has been ordered in the United States without later being canceled since 1973.

Perhaps the greatest irony of the history of nuclear power in the United States in the early 1980s was the effect of TMI on the nuclear industry. That single event probably accomplished more to interrupt the growth of nuclear power in this country than all of the activities of opposition groups over the preceding two decades. By the time of the Chernobyl disaster in 1986, Americans had decided—consciously or not—to accept the substantial and well-documented health and environmental risks of fossil-fueled power plants in preference to what they seemingly regarded as the far greater threats posed by nuclear power.

NUCLEAR WASTE ISSUES

Plant safety was by no means the only problem facing the nuclear power industry in the mid-1980s. Indeed, another issue—that of nuclear wastes—was gradually becoming at least as troubling as those surrounding the construction and operation of a nuclear reactor. As with any industrial operation, the use of nuclear materials for power production, scientific research, medical tests and treatments, and other applications generates waste materials that must be disposed of, preferably in some environmentally sensitive way. For example, when uranium is mined, very large amounts of rock, dirt, and other waste materials are left behind. These waste materials are called *mill tailings*, or simply *tailings*. At one time, these wastes were used for a variety of construction purposes, such as landfill, as the sand component of concrete that was then used to build roads, walks, drives, and concrete block, and in brick mortar. However, as the dangers posed by these materials became more obvious, the tendency to use tailings in any settings where humans might come into contact with them decreased.

As of 1999, the U.S. Agency for Toxic Substances and Disease Registry estimated that there were about 140 million tons of uranium tailings in the United States. In the vast majority of cases, these wastes are simply left in large storage ponds near the site from which they are extracted. The primary risk posed to human health by these tailings is lung cancer that results from the inhalation of radioactive gases released by the tailings.

HIGH-LEVEL WASTES

Another source of nuclear waste is the spent fuel from nuclear reactors. *Spent fuel* is the term used to describe uranium and plutonium removed from a reactor at a point that it can no longer sustain a nuclear chain reaction. Spent fuel is very hot, in both the thermal and radioactive sense. On average, about one-fourth to one-third of the fuel assemblies in a reactor core are replaced each year. This spent fuel is then stored at the reactor site in one of two kinds of locations: in spent fuel pools or dry cask storage sites. Spent fuel pools are large, swimming pool–like repositories in which fuel assemblies are kept under at least six meters (20 feet) of water.

A second method of storing spent fuel is by enclosing it in steel drums or steel cylinders filled with an inert gas. The drums and cylinders are designed to be both air-tight and leak-proof, although experience has shown that such cylinders have a tendency to crack and break open, releasing radioactive materials into the surrounding environment. Dry-cask storage has been used both for wastes removed directly from a reactor core and for wastes that have had a chance to cool off in a spent fuel pool, a process that often takes a few dozen years or so.

As of January 2004, the last date for which information is available, 38,413.7 metric tons (87,806.7 short tons) of uranium from 135,972 spent fuel assemblies had been removed from nuclear power plants in the United States. Of this amount, about 98 percent was being stored at the same location as the power plant from which it had been removed, with the remaining 2 percent having been removed to other storage locations.

Spent fuel is one form of high-level nuclear waste, a term that refers to nuclear waste that has two primary characteristics: (1) It is very radioactive now and (2) it will continue to be very radioactive for very long periods (thousands or tens of thousands of years into the future). A second kind of high-level nuclear waste is defense high-level nuclear waste, resulting from programs for the production of nuclear weapons. High-level nuclear waste consists of two kinds of isotopes: (1) those with short half lives that emit high levels of radiation but tend to become less dangerous within a relatively short period of time (a few hundred years), and (2) those with long half-lives that emit somewhat lower levels of radiation but tend to pose a risk for a much longer period of time, usually many tens of thousands of years. The half-life of a radioactive isotope is the time it takes for half of the isotope to decay, that is to give off radiation and change into a different isotope.

The two isotopes of uranium present in high-level nuclear wastes, uranium-235 and uranium-238, have half-lives of 713 million years and 4.5 billion years, respectively. Clearly, both isotopes pose a threat to human health for an extended period of time once they have been removed from a reactor core.

LOW-LEVEL WASTES

By far the largest single volume of nuclear waste is low-level waste. Low-level nuclear waste is defined, by default, as any type of nuclear waste that is not classified as high-level or transuranic waste (defined and discussed below). By that definition, low-level wastes emit significantly less intense radiation over shorter periods of time than is the case with high-level wastes.

About 90 percent of all the nuclear wastes generated in the United States are low-level wastes. About two-thirds of those wastes are produced from weapons research and development, and about one-third from commercial operations. Spent fuel contributes a very small amount to the overall volume of nuclear wastes produced, about 0.2 percent. The volume of wastes produced is less significant, however, than the amount of radioactivity contained in the wastes. Spent fuels tend to be highly radioactive, accounting for nearly all (96 percent) of the radioactivity produced by nuclear wastes in this country.

Low-level wastes are generated at nuclear power plants, fuel fabrication facilities, uranium reprocessing plants, hospitals, medical schools, universities, chemical and pharmaceutical manufacturers, and research laboratories. They consist of the everyday waste materials used at these facilities, materials such as scrap paper, rags, plastic bags, masks, gloves, protective clothing, cardboard, packaging material, organic fluids, and materials used in water-treatment systems.

During the early years of the so-called nuclear age, in the 1940s and 1950s, low-level wastes were usually disposed of at the locations where they were generated by placing them inside a concrete-filled barrel, which was then buried underground or, in some cases, dumped into the ocean. This solution was entirely unsatisfactory, however, as the concrete and barrels had a tendency to crack open, releasing radioactive materials into the ground or into ocean water.

By the late 1970s, most low-level wastes were being shipped to one of three central burial sites, at Beatty, Nevada; Richland, Washington; and Barnwell, South Carolina. By the end of that decade, however, the governors of Nevada and Washington had decided that they no longer wanted to be the dumping ground for low-level nuclear wastes for the rest of the country. They told the federal government that some national plan would be needed to force all 50 states to have a share in the disposal of these dangerous materials, and they temporarily closed their dumps to wastes coming from outside their states.

In response to this dilemma, the U.S. Congress, in 1980, passed the Low-Level Radioactive Waste Policy Act, requiring every state to become responsible for all the low-level nuclear wastes generated within its borders. The act allowed states to act on their own but also encouraged them to join together to form interstate compacts. In such cases, a single disposal site would be selected for each of the compacts, a site to which all nuclear wastes in the member states could be shipped.

The act required that states select some method for the disposal of low-level wastes by January 1, 1986, but, as that date approached, it became clear that only Washington, Nevada, and South Carolina were going to meet the deadline. In response to this situation, Congress passed the Low-Level Radioactive Waste Policy Amendments Act of 1985, providing for sanctions against states that did not make progress toward the development of their own waste disposal site or were not involved in a joint compact with other states to do so. The act set six "milestones" that states were required to meet in 1986, 1988, 1990, 1992, 1993, and 1996 in achieving compliance with the act.

As of 2004, 10 compacts had been created, including all 50 states except for Massachusetts, Michigan, New Hampshire, New York, North Carolina, and Rhode Island. None of the compacts had, as yet, opened a site for the disposal of low-level wastes, some had not yet selected a site, and some appeared to be questioning their commitment to Congress's plan for low-level waste disposal. For example, the Midwest Compact expelled Michigan from its union in 1991 when it became apparent that the state had no intention of honoring its role as "host" for the compact's nuclear wastes. The composition of some compacts had shifted over time also, with some unusual geographic associations resulting from those shifts. The newest compact, the Texas Compact, for example, consists of Texas and two New England states, Maine and Vermont.

TRANSURANIC WASTES

A third major category of wastes is called transuranic wastes. Transuranic wastes are similar to low-level wastes, except that they include a significant amount of radioactive isotopes with very long half-lives (a few thousand years or more). Transuranic wastes are produced almost exclusively as a result of nuclear weapons research and development. They pose a somewhat different disposal problem than do low-level wastes since the threat they pose to human health and the environment is likely to endure for a very long time, at least a few thousand years. Prior to 1970, transuranic wastes were stored in essentially the same way as other low-level wastes. They were imbedded in concrete-filled steel drums and then buried in shallow ditches. Scientists and politicians alike were aware that this method of disposal was entirely unsatisfactory because of the long-term threat such wastes posed to human health and the environment. It was not until 1979, however, that the U.S. Congress was able to devise a plan for the storage of these wastes. In that year, Congress passed the Defense Authorization Act of 1979 (now Public Law 96–164) authorizing the construction of a permanent underground storage area for transuranic wastes to be called the Waste Isolation Pilot Plant (WIPP). The site selected for the storage area was located in the Chihuahuan Desert of Southeastern New Mexico. The giant cave constructed to hold the wastes was built about 650 meters (2,150 feet) underground inside a 600-meter (2,000-foot) thick salt formation. Geological studies indicate that the formation has been stable for more than 200 million years, providing the kind of time frame needed for long-term storage of transuranic wastes.

Selection of the site and construction of the facility was hampered, not surprisingly, by objections of some New Mexico residents and environmental groups who were not convinced that the storage area was as safe as claimed. In spite of those problems, however, the project went forward with relatively few delays, and WIPP received its first shipment of transuranic wastes on March 26, 1999. Over the next three decades, the facility is expected to receive 19,000 shipments of transuranic wastes from 23 locations nationwide where they are now stored.

THE YUCCA MOUNTAIN SAGA

Of all the problems facing the nuclear power industry, none has been so difficult to resolve as the fate of the high-level wastes produced by nuclear power plants and weapons research and development every year. The U.S. Department of Energy (DOE) has estimated that, as of 2004, nuclear power plants in the United States had generated about 49,000 metric tons (54,000 short tons) of high-level radioactive wastes. These wastes are currently being stored in on-site temporary facilities in 68 locations around the country. The total amount of radioactivity contained within these wastes has been estimated at about 30 billion curies, with perhaps 300 times that

47

amount coming from high-level defense wastes. DOE estimates, further, that the amount of nuclear waste that must be disposed of will rise to a total of about 105,000 metric tons (116,000 short tons) by the year 2035.

Scientists and politicians realized early on in the nuclear age that some means would have to be found to dispose of the highly radioactive, long-lived wastes produced by both weapons projects and electrical power generation. As early as 1955, the Atomic Energy Commission (AEC) asked the National Academy of Sciences (NAS) to undertake a study of this problem. Two years later, the NAS issued a report that recommended burying nuclear wastes deep within geological formations known as salt domes. Salt domes have the advantage of being largely impermeable to water and resistant to heat, greatly reducing the likelihood that radioactive materials could seep out into the surrounding groundwater. Salt domes also have a tendency to heal themselves when cracks develop within them.

The first potential site to which the AEC turned its attention was the Salina Basin salt bed that lies underneath large parts of Michigan and Ohio. That line of research came to an end, however, when officials in the two states heard about the nuclear waste project being considered for their area and demanded that the AEC terminate its studies of the Salina Basin. AEC researchers then turned their attention to another large salt bed located in the Midwest, focusing specifically on the region near Lyons, Kansas. By 1970, they had become convinced that a large salt dome in this area would provide a satisfactory location for the nation's first (and probably only) high-level nuclear waste depository. That decision remained in force for only two years, however, as further research showed that drilling in the Lyons salt dome caused enough damage to the dome that its long-term integrity could not be guaranteed. The AEC responded to this news by announcing that high-level wastes would have to continue to be stored at the sites where they were generated for the foreseeable future.

During the rest of the 1970s, the Energy Research and Development Agency continued studying salt domes in 36 states, looking for a satisfactory disposal site. Neither the federal government nor private industry was willing to provide the funds necessary to support this research, however, and little progress was made in locating a final resting place for the nation's high-level wastes. Then, in 1982, the U.S. Congress (nearly 40 years after the beginning of the nuclear age) passed legislation designed to deal finally and conclusively with this problem. It adopted the Nuclear Waste Policy Act (NWPA) that declared that (1) the nation's policy on nuclear wastes would be for their disposal in secure geological formations, (2) the federal government would be responsible for managing and disposing of high-level nuclear wastes, (3) the Department of Energy (DOE) would be responsible for the design, construction, and operation of the geological depositories, (4) the

Environmental Protection Agency (EPA) would be responsible for establishing and enforcing environmental standards, and (5) the Nuclear Regulatory Agency (NRA; a second successor agency of the AEC) would be responsible for licensing and overseeing the operation of the nuclear repository.

The 1982 NWPA legislation also required the creation of two distinct repository sites so that all the nation's nuclear wastes would not be allocated to a single part of the country. It set a deadline of 1998 for the opening of at least one of those sites and created a Nuclear Waste Fund to pay for the development of the waste sites. The fund was to be paid for by a tax assessed on companies that operate nuclear power plants.

By 1983, DOE had chosen nine sites in six states as possible repository sites. The six states were Louisiana, Mississippi, Nevada, Texas, Utah, and Washington. Three years later, the secretary of energy had winnowed down that list to five sites, one each in Mississippi, Nevada, Texas, Utah, and Washington, and President Ronald Reagan selected three of those sites for further study. Those sites were Hanford, Washington; Deaf Smith County, Texas; and Yucca Mountain, Nevada. A year later, Congress passed an amendment to the NWPA requiring the DOE to focus exclusively on the Yucca Mountain site. That decision was based in part on a desire to reduce additional costs of developing a nuclear waste facility and, to some extent, a concern about the ongoing delay in making a decision on this issue.

A year after adoption of the NWPA amendments, the secretary of energy announced that his department would not be able to meet the original legislation's requirement of having a repository in place by 1998, and that 2003 would be a more likely date for that to happen. Over the next decade, however, progress on the Yucca Mountain site was delayed for a number of reasons, one of the most important of which was Nevada's very strong objection to the siting of the facility at Yucca Mountain. In addition, the project was besieged by a number of technical problems, and costs continued to mount. Although initial estimates for the cost of the project was set at somewhat less than $1 billion, preliminary costs as of 1996 had already reached more than $4 billion.

For most of the 1990s, proponents and opponents of the Yucca Mountain project carried out a host of research projects attempting to design a safe facility and to demonstrate that it would, in fact, be safe (or, in the case of opponents, that it could *not* be safe). Very slowly, the bureaucratic hurdles delaying completion of the project were overcome. In 1997, for example, the DOE completed construction of an Exploratory Studies Facility at the Yucca Mountain site, a facility designed to test the technology that was ultimately to be employed at the site. In 1998, DOE issued a Yucca Mountain Viability Assessment that detailed the steps that were still necessary to construct the facility, described its operation, outlined the procedures needed to obtain licensing of the repository, and estimated its final cost. A year later, the EPA issued its environmental impact statement for the site, outlining the necessary radiation protection steps needed to ensure safety of humans in the area and the environment. Optimists began to think that the nation would soon have its long-awaited repository for high-level nuclear wastes.

Such beliefs turned out to be wishful thinking. By 2004, construction on the waste site had not begun. Indeed, DOE had not yet even requested approval from NRC to begin construction at the location. New problems continued to arise that fed concerns about the safety of storing high-level wastes deep within the mountain. In late 2003, for example, DOE studies showed that groundwater leaking into the repository site might cause deterioration of the canisters in which nuclear wastes were to be buried, causing them to leak and release radioactive materials into the groundwater. In August 2004, the Federal Appeals Court for the District of Columbia ruled that DOE's plans for ensuring the safety of stored wastes for a period of 10,000 years was inadequate. The department needed to amend its plan, the court said, to provide for an even longer period during which the general public would be protected from radiation although it did not specify, precisely how long that period had to be.

In spite of these setbacks, DOE still seems to believe it can open the Yucca Mountain site by 2010, although outside authorities believe this estimate is overly optimistic. Should the facility *not* be ready by 2010, the nation would enter the second decade of the 21st century, nearly 70 years after the first use of nuclear energy, without a final resting place for the thousands of tons of radioactive wastes still sitting in "temporary" sites throughout the United States.

The design of the Yucca Mountain repository site has changed in some respects over the two decades during which it has been considered. In general, the plan is to bore at least 50 tunnels into the mountain at levels of between 200 and 500 meters (660 and 1,600 feet) beneath the surface, and about 300 meters (1,000 feet) above the water table, the level below which the soil remains saturated with water. The total length of all the tunnels is estimated to be about 150 kilometers (100 miles). Radioactive wastes transported to the site from around the country will be delivered first to one of 29 buildings where they will be encased in specially designed canisters made of very strong, highly resistant nickel alloy. The canisters will then be delivered by remote-control equipment from the surface buildings to the tunnels, where they will be placed under drip shields made of a titanium alloy designed to protect them from water seeping into the tunnel and dripping from the ceiling. As each tunnel fills, it will be sealed off with specially designed barriers that allow the release of heat generated by the wastes but protect against the escape of radiation into the surrounding area.

DECOMMISSIONING NUCLEAR POWER PLANTS

The future of nuclear power production in the United States includes yet one more difficult problem: the decommissioning of old nuclear power plants. As with any other industrial facility, nuclear power plants grow old, become outmoded, and must eventually be shut down and dismantled. The life expectancy for a nuclear power plant is about 30 years, after which time it must be decommissioned. A 30-year limit on the operation of nuclear power plants is, in fact, included in all licenses issued by the NRC.

The process of decommissioning a nuclear power plant is monitored by the Nuclear Regulatory Commission. When a company decides to close down a plant, it applies to the NRC for permission to proceed with decommissioning. Three forms of decommissioning are available: DECON, SAFSTOR, and ENTOMB. A DECON permit allows the company to close the power plant, remove equipment, tear down the structure, and transfer wastes and contaminated materials to a safe storage site. SAFSTOR simply adds one step to this process, allowing a company to stop power production at a plant but permitting it to maintain the plant until some future date, at which time DECON can begin. ENTOMB is a procedure used when unusually high levels of radiation are present at a site. The plant is encased in a structurally sound material to prevent the release of radiation to the surrounding environment. The site is then monitored and maintained until it is safe to begin a DECON procedure.

The cost of decommissioning a nuclear power plant depends on a number of factors, primarily the type of reactor used and the geographic location of the plant. Currently, the average cost of decommissioning a plant is about $325 million. This cost is paid for by the nuclear industry, which collects a tax from ratepayers of 0.1–0.2 cents per kilowatt hour of electricity generated. These funds are deposited in a trust fund managed by the Nuclear Regulatory Commission for use in paying off future costs of decommissioning power plants. As of 2001, the last year for which data are available, the trust fund had assets of about $22.5 billion in anticipation of future costs of about $40 billion for decommissioning of all existing power plants in the United States.

As with almost any issue related to nuclear energy, the decommissioning of nuclear power plants is viewed with considerable concern by many individuals. They point out the risks associated with the dismantling, transport, treatment, and storage of wastes, equipment, fuel, and other materials taken from a plant that has been shut down. It is impossible, they say, to prevent some radioactive materials from remaining at a plant site or escaping into the surrounding environment, thereby increasing the risk of illness and death to humans, other animals, and plants. The risks associated with decommissioning of nuclear reactors are only one more reason, they suggest, that such plants should never be built in the first place.

NUCLEAR TERRORISM

On September 11, 2001, 19 members of the al-Qaeda terrorist network hijacked four commercial airliners flying out of Newark International, Washington Dulles, and Boston's Logan airports. They crashed three of those airliners into the twin towers of New York's World Trade Center and the Pentagon, in northern Virginia. The fourth plane was forced to crash in the Pennsylvania countryside by the heroic actions of passengers on the flight. A secondary consequence of this event, the worst attack on American citizens to occur on U.S. soil, was a renewed concern among politicians, scientists, bureaucrats, and ordinary citizens about the dangers of nuclear terrorism. Two of the hijacked planes had flown less than 30 miles from the Indian Point nuclear power facility in Buchanan, New York. The Three Mile Island nuclear power facility near Harrisburg, Pennsylvania, was not far from the crash site of the fourth aircraft and might, according to some observers, have even been the target of that plane's hijackers.

The term *nuclear terrorism* refers to the use of nuclear materials—enriched uranium or plutonium in fuel rods, spent fuel, nuclear wastes, and other materials—in a terrorist attack. Such materials could conceivably be used in the construction of a nuclear weapon, although the technology required for such a project is quite sophisticated and probably beyond the ability of most terrorist groups. Far more likely would be the use of nuclear materials for the construction of a dirty bomb, also known as a radiation dispersal device (RDD). RDDs are conventional bombs to which have been added radioactive materials. When such devices are exploded, they release high levels of radiation over wide areas. The construction of a dirty bomb is well within the capability of virtually any terrorist group, provided it can gain access to nuclear materials.

Direct attacks on nuclear facilities by terrorists are another form of nuclear terrorism. Such attacks might take the form, for example, of a highjacked aircraft flown into a nuclear facility, similar to the attacks on the World Trade Center and the Pentagon; a car or truck filled with explosives driven into a nuclear power plant; or an armed attack by a group of terrorists on a nuclear facility.

In the first decade of nuclear power plant construction, the AEC did not view attacks such as these as credible threats to nuclear facilities. Licensing regulations specifically excused plants from preparing themselves against outside attacks. A portion of AEC regulations dealing with licensing requirements originally adopted on September 26, 1967, for example, said that "An applicant for a license to construct and operate a production or utilization facility . . . is not required to provide for design features or other measures for the specific purpose of protection against the effects of (a) attacks and destructive acts, including sabotage, directed against the facility by an enemy of the United States, whether a foreign government or other person."[10]

This position might have been defensible in 1967, one could argue, when the risk of highly destructive assaults on U.S. nuclear power plants might have seemed very remote. Within a decade, however, AEC's successor, the NRC, had changed its stance on the threat of sabotage. In 1977, the agency adopted regulations requiring nuclear facilities to have a security plan to protect against "radiological sabotage," described as "any act that could directly or indirectly endanger public health and safety by exposure to radiation."[11]

The guiding principle introduced by the NRC in these regulations was the design basis threat (DBT), defined as "the maximum terrorist threat that a facility must prepare to defend against."[12] The DBT contained a number of elements describing the nature of an attack against which a nuclear power plant was required to defend itself. Those elements included (1) an attacking force of "several" (later defined as "three") well-trained individuals (2) armed with weapons no larger than hand-held automatic weapons (3) with the assistance of no more than one inside collaborator (4) operating from a vehicle no larger than a four-wheel-drive land vehicle.

Some observers argued that these provisions were inadequate and that nuclear power plants should be required to have plans to protect themselves against attacks by larger groups of individuals far better equipped to inflict damage on the plant. For example, the International Task Force on Prevention of Nuclear Terrorism (ITFPNT) convened by the Nuclear Control Institute and the Institute for Studies in International Terrorism of the State University of New York in 1986 warned that "the probability of nuclear terrorism is increasing." It pointed out that NRC's design basis threat model had "several problems." "The models were designed a decade ago when the threat of nuclear terrorism was thought to be mostly from anti-nuclear protesters. Today's wide range of threats is not covered by the models."[13]

The NRC largely ignored those suggestions until a truck bomb was exploded in the basement of the World Trade Center (WTC) on February 26, 1993. At that point, the agency made its first changes in its DBT plans in 16 years. Those changes called for increases in plant safety requirements to include attacks by truck bombs, like that which had occurred at the WTC. Except for that change, however, the 1977 DBT requirements remained largely in place.

NRC monitors the effectiveness of its DBT regulations through a subsidiary program known as the Operational Safeguards Response Evaluation (OSRE) program. OSRE arranges for so-called force-on-force or black hat tests during which it assesses a plant's ability to withstand a terrorist attack. The OSRE tests are normally announced six months in advance, at which time the plant is notified of the time of the black-hat event. In spite of the generous conditions of such tests, they have not resulted in very promising results. In the year starting May 2000, for example, 11 black-hat tests were conducted, only two of which resulted in a plant's successfully repelling the intended invaders. Of the nine plants that failed the test, two reacted successfully in the respect that the plant would not have been destroyed. Seven of the nine plants, however, were unable to carry out actions necessary to prevent destruction of the plant and meltdown of the reactor core. As one critic of the NRC said, "This is a disgraceful situation that no amount of spin control by the NRC and the nuclear power industry can hide."[14]

Even more troublesome about these results was the fact that they were conducted as part of the NRC's voluntary inspection system, known originally as its Self-Assessment Program (SAP; later changed to Safeguards Performance Assessment, SPA). Thus, the industry had problems in regulating itself vis-a-vis safety against terrorist attacks even when plants knew months in advance they were to be visited.

As one would have hoped, NRC responded quickly to the terrorist attacks of September 11, 2001 (9/11). It ordered plants to go to the highest level of security available, issued recommendations and orders for increasing security arrangements, and began a reassessment and revision of the existing DBT. At the same time, representatives of both the NRC and the nuclear industry rushed to assure the public that U.S. nuclear power plants were safe against the kind of attacks by jet airplanes that terrorists had used on 9/11. For example, NRC chairman Richard Meserve told the National Press Club on January 17, 2002, that "the physical protection at nuclear power plants is very strong. . . . The plants are among the most formidable structures in existence and they are guarded by well-trained and well-armed security forces. The security at nuclear plants has always been far more substantial than that at other civilian facilities and it has been augmented since September 11th."[15] Meserve's confidence was echoed by a number of industry spokespersons. For example, Lynette Hendricks of the Nuclear Energy Institute announced less than a week after the 9/11 attacks that "We believe the plants are overly defended at a level that is not at all commensurate with the risk.[16]

NRC and NEI representatives have been especially concerned about assuring the public that nuclear power plants are safe against attack by jumbo jets like the Boeing 757 and 767 aircraft used on 9/11. On the same day these attacks occurred, NRC spokesperson Brock Henderson announced that "containment structures are designed to withstand the impact of a 747." Ten days later, Henderson corrected himself, admitting that nuclear power plants are not designed to withstand direct hits by the large commercial airliners now in use. "The initial cut we had on that (issue) was misleading," he explained.[17]

This assessment corresponds more closely with the views of independent experts on nuclear energy. David Kyd, spokesperson for the International Atomic Energy Agency, for example, has pointed out that most nuclear power plants were built prior to the advent of large modern jumbo jets. "If you postulate the risk of a jumbo jet full of fuel, it is clear that their design was not conceived to withstand such an impact."[18] Indeed, industry officials and regulators now admit that protecting nuclear power plants against such attacks is "beyond the capabilities of private security forces" and that this level of security would "be one for the [U.S.] military."[19]

Still, NRC officials and industry representatives seem convinced that they have responded well to the events of 9/11 and have brought security at nuclear power plants to the highest possible level. For example, A. D. Barginere, lead security specialist at Progress Energy, operators of four nuclear power plants in the Carolinas and Florida, has said that "We are held to a higher standard by the Nuclear Regulatory Commission. We believe and support that standard. We implement security and take it very seriously at the hardened stations that we are." As a result, according to Barginere, "Nuclear security should be the role model, should be the standard of security in America."[20]

Independent observers and critics of nuclear power plant security do not necessarily agree with the optimism expressed by NRC and the nuclear industry. For example, an extensive study on efforts to improve security at nuclear power plants since 9/11 by the Government Accountability Office (GAO) found that "NRC responded quickly to the September 11, 2001, terrorist attacks with multiple steps to enhance security at commercial nuclear power plants."[21] But the study also found that a great deal still needs to be done to ensure the safety of the nation's nuclear power plants against terrorist attacks. The two major problems facing NRC, according to the study, are that the NRC's review of plant security plans "has been rushed and is largely a paper review" and that the agency's force-on-force exercises needed to test those plans will take up to three years.[22] Even then, the effectiveness of these exercises is likely to be questionable, since NRC is likely to allow plants to use attack forces of their own choosing.

Meanwhile, evidence continues to accumulate that the integrity of nuclear power plant security may be somewhat less robust than the NRC would hope and that industry claims. For example, the nonprofit organization Project on Government Oversight (POGO) conducted a survey in

2002 of "more than 20" security guards working at 24 nuclear reactors in 13 nuclear power plants. It asked these guards how effectively their facility would be able to repel an attack by terrorists. About one-quarter of the guards expressed "confidence" that the plant could defeat a terrorist attack. The remaining interviewees were doubtful about the plant's ability to stave off such an attack. For example, one respondent, identified as "Guard F," said that "The NEI (Nuclear Energy Institute) is fooling the public, which is outrageous." Another respondent, identified only as "Guard A," claims that "Morale sucks" at his facility. And he believes that the efforts of the NRC leave much to be desired. "They're more of a cheerleader for the nuclear industry than a watchdog," he suggests.[23]

More than four years after the terrorist attacks of 9/11, then, experts continue to dispute the security of nuclear power plants. On the one hand, the industry argues that "No other private industrial facilities have the combination of robust physical protection, well-trained and armed security forces and emergency response capability that is found at every nuclear power plant in the United States."[24] On the other hand, critics continue to believe that "despite September 11—when the NRC's assumptions crumbled at the moment the Twin Towers fell—both the industry and the agency that regulates it continue to resist making any significant improvement to dismally inadequate and outmoded security regulations" and that the NRC is "behind the curve, 'fighting the last war' rather than protecting against threats that can materialize without warning."[25] One can only hope it will not take another terrorist attack to decide which of these views is correct.

NUCLEAR RENAISSANCE

With the election of President George W. Bush in 2000, the long decline in nuclear power plant development in the United States appeared to be over. Indeed, prospects for the construction of new reactors seemed so good that a number of enthusiasts began to talk about a nuclear renaissance, both in the United States and around the world. Beginning in 2000, virtually every major conference dealing with nuclear power plant issues contained at least one session with the phrase "nuclear renaissance" in its title. One of the most prominent of these conferences was a meeting held in Washington, D.C., on September 10–12, 2002, at which industry representatives and representatives of the U.S. government optimistically surveyed the outlook for nuclear power in the coming generation. The following spring, a Nuclear Renaissance Forum was held in Chicago under the sponsorship of Framatone and Westinghouse, two nuclear plant manufacturers.

Some observers pointed to the fact that a nuclear renaissance had already begun. In the first place, increase in the output from existing nuclear plants had mushroomed in the decade of 1990s by an amount equivalent to that which could have been provided by 23 new 1,000 megawatt plants. This increase had been made possible by increasing efficiency of existing plants. The share of U.S. electrical energy provided by nuclear power had increased to 20 percent by 2000. And, of considerable importance, public opinion about nuclear power plants seemed to have taken a more positive direction. In a study conducted by the Bisconti Research company in 2004, for example, 65 percent of those interviewed responded positively to the question, "Overall, do you strongly favor, somewhat favor, somewhat oppose, or strongly oppose the use of nuclear energy as one of the way to provide electricity in the United States?" That number was the highest found by researchers in surveys going back to 1983. At that point, only 46 percent of those interviewed responded positively to this question, while 49 percent expressed opposition to the development of nuclear power. Interestingly enough, and for whatever it is worth, the opinion spread was greatest among college graduates who favored the development of nuclear power plants by a margin of 73 percent to 24 percent (with 3 percent having no opinion).

A nuclear renaissance is likely to be even more dramatic on an international scale than it would be in the United States. The International Atomic Energy Agency has predicted that demand for electricity is likely to increase fivefold by 2050, requiring a quadrupling of the number of nuclear power plants operating by then. The greatest increase in demand for electricity and nuclear capacity is likely to occur in developing nations, especially those in the Far East. China has announced it will quadruple its nuclear capacity (by adding 32 new reactors to its existing 11 plants) by 2020. And India expects to increase its capacity from 14 plants in 2004 to more than 40 plants in 2012.

Analysts have pointed to a number of factors that has led to a nuclear renaissance. One important factor has been increased safety at nuclear power plants in the United States. No major accident has occurred in this country since the Three Mile Island event in 1979, and the public now appears to look more favorably on the safety of nuclear power production. In the 2004 Bisconti survey, 60 percent of interviewees gave nuclear power plants a "safe" rating, while 19 percent thought of them as being "unsafe."

A second factor in promoting a nuclear renaissance has been a significantly more favorable regulatory atmosphere in the federal government. President Bush's National Energy Policy, promulgated in May 2001, strongly recommended increased attention to nuclear power as a source of energy in the nation's future plans. It made a number of specific recommendations designed to achieve this objective, including encouraging the NRC to facilitate industry efforts to expand the number of nuclear power plants in operation in the United States, urging the commission to extend licenses on existing plants and suggesting the need for legislation that would transfer some of the financial risk of nuclear power plant construction and operation from industry to the federal government.[26]

Administration officials from the cabinet level down made a number of speeches throughout the nation and around the world reinforcing these themes and emphasizing how important the federal government regarded the growth of nuclear power here and throughout the world. In addition, members of Congress sympathetic to the president's views on nuclear power began to introduce a number of bills that would promote the development of nuclear power. One such bill, for example, eventually led to the adoption of a Strategic Plan for Light Water Reactors Research and Development that provided federal support for industrial research on new reactor designs.[27]

This new perspective on nuclear power was also expressed in a relatively short period of time by regulatory agencies. Indeed, one spokesperson for industry praised the new outlook at NRC that had appeared to be evolving in the early 21st century. "For many years, the Nuclear Regulatory Commission (NRC) regulatory process was unstable," he said. "Beginning a few years ago, the NRC, with the support and assistance of the industry, embarked on a program of reform designed to be more objective, more focused on safety significant matters, and reflecting a risk-informed philosophy. As a result of these initiatives, the regulatory process is much more predictable, thereby reducing investor uncertainty."[28]

Yet another factor involved in the nuclear renaissance has involved environmental concerns. As the nation and the world grow more concerned about the deleterious effects of waste gases produced by fossil fuel plants, the relatively pollution-free advantages of nuclear power generation have become more and more important. Especially after the signing of the Kyoto Protocol on Climate Change in 1992 (which did not include the United States), various industries have been exploring methods for the production of energy with reduced emissions of greenhouse gases, such as the carbon dioxide produced during combustion of fossil fuels. More and more, industry representatives are asking opponents of nuclear power production to weigh the possible safety risks of nuclear plants (which they insist are vanishingly small) with the very considerable and very real environmental hazards posed by the continued use of fossil fuel combustion to produce electricity.

Finally, renewed interest in the development of nuclear power plants results from the economic advantages of nuclear power production. In 2002, the cost of electricity generated by nuclear power plants was 1.71¢/kwh (cents per kilowatt-hour). By comparison, electricity from coal-powered plants cost 1.85¢/kwh; from oil-powered plants, 4.41¢/kwh; and from gas-powered plants, 4.06¢/kwh. Given this clear advantage of nuclear plants, supporters say, greater emphasis should be placed on reactors as a means of solving the nation's growing need for electricity in the years to come.

Enthusiasm for a nuclear renaissance is far from unanimous. Many individuals and organizations that have been battling the use of nuclear power for decades deride the case made by the administration and by the nuclear power industry. They prefer to describe current trends as "nuclear fantasy," "nuclear nightmare," or "nuclear resurgence," a revival of most of the same problems with nuclear reactors that they have been criticizing for half a century. They argue that George W. Bush, whom they sometimes call "The Nuclear President," is pushing a nuclear agenda at least partly to benefit his own friends and supporters in the energy industry.

One aspect of a nuclear renaissance that troubles some observers is the continued threat of nuclear terrorism. As more nuclear power plants come on line in the United States and around the world, more targets become available to terrorists, and more difficult become the problems of providing security for nuclear reactors. Alain Marsaud, president of the domestic security group in the French Parliament, has highlighted this issue: "[A nuclear renaissance is] really bad timing. We're coming to the end of the economic use of fossil fuels at a time when terrorists are trying to get their hands on nuclear material or target nuclear infrastructure. If the world is condemned to use more nuclear power it will be a real challenge."[29]

So, as enthusiasm for the promise of nuclear power grows in some parts of the nation and many parts of the world, the general public is once again treated to a rosy view of the future in which nuclear fission will benefit men and women throughout the world. At the same time, however, critics continue to raise issues about safety, cost, environmental effects, and other aspects of nuclear power that, they say, will prevent it from ever reaching the potential promised by supporters. Fundamental questions about the role of nuclear energy in the life of everyday individuals, first raised in the late 1940s, are still not completely answered. They are likely to remain to challenge the world for the foreseeable future.

---

[1] James Bryant Conant, quoted in "The Italian Navigator Has Landed in the New World." Secret Race Won with Chicago's Chain Reaction. Available online. URL: http://www.lanl.gov/worldview/welcome/history/08_chicago-reactor.html. Downloaded on February 8, 2005.

[2] Lewis L. Strauss, quoted in "Abundant Power from Atom Seen," *The New York Times*, September 17, 1954, p. 5.

[3] Dwight D. Eisenhower, "'Atoms for Peace' Address to the United Nations General Assembly, December 8, 1953," in Philip L. Cantelon, Richard G. Hewlett, and Robert C. Williams, *The American Atom: A Documentary History of Nuclear*

Nuclear Power

Policies from the Discovery of Fission to the Present, Second edition. Philadelphia: University of Pennsylvania Press, 1991, pp. 103–104.

[4] Gunter G. Pretzsch, et al., "The Chernobyl Sarcophagus Project of the French-German Initiative: Achievements and Prospects," paper presented at the 22nd Annual ESRI International User Conference, San Diego, California, July 8–12, 2002, paper 658. Available online. URL: http://gis.esri.com/library/userconf/proc02/pap0658/p0658.htm. Downloaded on February 8, 2005.

[5] David Brower, quoted in McGee Young, "Creating a Modern Lobby: The Sierra Club and the Rise of Environmentalism, 1892–1970," a paper presented at the Annual Fellows Conference of the Miller Center of Public Affairs, University of Virginia, May 9–10, 2003, Charlottesville, Virginia. Available online. URL: http://www.americanpoliticaldevelopment.org/townsquare/print_res/in_progress/young.pdf. Downloaded on February 8, 2005.

[6] Quoted in J. A. L. Robertson, "A Brief on (A) Canada's Domestic Nuclear Issues & (B) International Nuclear Trade," a paper submitted to the First and Second Weeks' Panels of the Interfaith Program for Public Awareness of Nuclear Issues, Toronto, September 24, 1984. Available online. URL: http://www.magma.ca/~jalrober/IPbriefa.htm. Downloaded on February 8, 2005.

[7] *Nuclear Power in an Age of Uncertainty*. Washington, D.C.: U.S. Congress, Office of Technology Assessment, OTA-E-216, February 1984, p. 211.

[8] "Statement of Purpose," Critical Mass Energy and Environment Program. Available online. URL: http://www.citizen.org/cmep/. Downloaded on February 8, 2005.

[9] Alan Weinberg, quoted in "The First 50 Years," *Oak Ridge National Laboratory Review*, 1992–1994. Available online. URL: http://www.ornl.gov/info/ornlreview/rev25-34/chapter6.shtml. Downloaded on February 8, 2005.

[10] Code of Federal Regulations, Chapter 10, Section 50.13.

[11] Code of Federal Regulations, Chapter 10, Section 73.2.

[12] Jim Wells, "Preliminary Observations on Efforts to Improve Security at Nuclear Power Plants," United States Government Accountability Office, Report GAO-04-1064T, September 2004, p. 1. Also available online. URL: http://www.gao.gov/new.items/d041064t.pdf. Downloaded on February 8, 2005.

[13] *International Task Force on Prevention of Nuclear Terrorism*. Washington, D.C.: The Nuclear Control Institute, June 25, 1986, p. 12. Also available online. URL: http://www.nci.org/PDF/NCI_6-25-1986_Report.pdf. Downloaded on February 8, 2005.

[14] Paul Leventhal, "Nuclear Power Reactors Are Inadequately Protected Against Terrorist Attack," statement before the House Committee on Energy and Commerce, Subcommittee on Oversight and Investigations, December 5, 2001. Available online. URL: http://www.nci.org/01nci/12/react-prot.htm. Downloaded on February 8, 2005.

[15] Richard A. Meserve, "Nuclear Security in the Post-September 11 Environment," Remarks to the National Press Club, January 17, 2002, Washington, D.C. Complete transcript available online at http://www.nrc.gov/reading-rm/doc-collections/commission/speeches/2002/s02-001.html. Downloaded on February 10, 2005.

[16] Lynette Hendricks, quoted in Douglas Pasternak, "A Nuclear Nightmare," *U.S. News & World Report*, September 17, 2001, p. 44.

[17] Brock Henderson, quoted in Daniel Hirsch, "The NRC: What, Me Worry?" *Bulletin of the Atomic Scientists*, January/February 2002, p. 38.

[18] David Kyd, quoted in Hirsch, p. 38.

[19] Meserve, *op. cit.*

[20] A. D. Barginere, quoted in Nuclear Energy Institute, "Transcript—Proven. Prepared. Protected. Security of America's Nuclear Power Plants." Available online. URL: http://www.nei.org/index.asp?catnum=3&catid=1139. Downloaded on February 8, 2005.

[21] Wells, p. 3.

[22] Ibid., p. 3.

[23] Quoted in Project on Government Oversight, *Nuclear Power Plant Security: Voices from Inside the Fences*, September 12, 2002. Available online. URL: http://www. pogo.org/p/environment/eo-020901-nukepower.html#ExecSum. Downloaded on February 8, 2005.

[24] Nuclear Energy Institute, "High-Caliber Security Forces Negate Need to Federalize Nuclear Plant Defenders, Report Shows," Press Release of January 8, 2002.

[25] Hirsch, p. 40.

[26] The full report can be viewed online at http://www.whitehouse.gov/energy/.

[27] The plan is available online at http://nuclear.gov/reports/LWR_SP_Feb04.pdf.

[28] Michael B. Sellman, "America's Nuclear Renaissance," Remarks made at the Ninth International Conference on Nuclear Engineering held in Nice, France, on April 12, 2001. Available online. URL: http://www.nmcco.com/newsroom/ presentations/anr.htm. Downloaded on February 8, 2005.

[29] Alain Marsaud, quoted in Katrin Bennhold, "Nuclear Energy Is Making a Comeback," *New York Times*, October 17, 2004, pp. A1, A4. Available online at http://www.nytimes.com/2004/10/17/international/europe/17CND-IHINUK. html?ex=1702395600&en=0619123baa8a877o&oi=5070.

CHAPTER 2

THE LAW AND NUCLEAR POWER

This chapter describes laws and legal decisions relating to the use of nuclear energy in the United States. Extracts from some of these laws and court decisions appear in the Appendices.

LAWS

The use of nuclear energy for both military and peacetime applications is strictly regulated by a number of federal, state, and local laws. This section focuses exclusively on laws relating to peaceful applications of nuclear energy, with special emphasis on those that determine the conditions under which nuclear power plants may be built, operated, and dismantled, as well as the environmental restrictions placed on such plants.

ATOMIC ENERGY ACT, PUBLIC LAW 79-585 (1946)

The Atomic Energy Act of 1946 was the first law passed relating to the control of nuclear energy in the United States. At the conclusion of World War II, the question arose as to the fate of the Manhattan Engineering District (MED), the program under which the first nuclear weapons were built. The preponderance of opinion among lawmakers appeared to be that the MED should be continued, probably in some modified form, such that control over the development and use of nuclear energy would remain in the hands of the U.S. military. In May 1945, Representative Andrew J. May (D-Ky.) and Senator E. C. Johnson (D-Col.) introduced a bill into the U.S. Congress aimed at such an objective. The May-Johnson bill would have assigned control over all nuclear energy development to the War Department.

At first, the bill seemed assured of passage, especially when President Harry S. Truman announced that he favored its provisions. The May-Johnson bill, however, aroused a considerable amount of concern among scientists, many of whom had worked on the Manhattan Project and appreciated the horror inherent in nuclear weapons. They were very troubled that the future development of nuclear energy would remain in the hands of the military solely. Although traditionally a largely nonpolitical group, the scientific community rapidly organized to oppose the May-Johnson bill and eventually threw its support to a competing bill introduced by Senator Brien McMahon (D-Conn.) in December 1945. The McMahon bill proposed the creation of a civilian agency that would take over responsibility for the development and promotion of nuclear energy in the United States. After a congressional debate that had lasted nearly a year, the McMahon bill was passed. It became the Atomic Energy Act of 1946.

The Atomic Energy Act of 1946 provided the fundamental structure under which nuclear energy was to be controlled in the United States, a structure that remains in place to the present day. The act authorized the establishment of two agencies, the Atomic Energy Commission (AEC) and the Joint Committee on Atomic Energy (JCAE) of the U.S. Congress. The role of the JCAE was to provide oversight on the activities of the AEC. The AEC was assigned six major responsibilities: (1) assisting and fostering private research and development of nuclear energy, (2) providing for the open and free dissemination of scientific information about nuclear energy, (3) sponsoring of research on nuclear energy, (4) controlling the production, ownership, and use of nuclear materials, (5) studying the social, political, and economic effects of nuclear energy, and (6) keeping Congress informed about developments in the field of nuclear science. ACE was also to be responsible for the design, development, construction, and maintenance of all nuclear weapons in the nation's arsenal. In order to carry out its functions, the Atomic Energy Commission was to be organized into four major sections: divisions of research, production, materials, and military applications.

In spite of its important accomplishments, the Atomic Energy Act of 1946 contained some serious defects. For example, an amendment offered by Senator Arthur Vandenberg (R-Mich.) gave veto power over AEC decisions to the committee's Board of Military Advisors, essentially limiting to some extent its scope of operations. Also, the monopoly on nuclear materials given to the committee by the act was a matter of serious concern to private industry, which had great hopes for the use of such materials in the development of many peacetime applications of nuclear science.

ATOMIC ENERGY ACT, PUBLIC LAW 83-703 (1954)

The Atomic Energy Act of 1954 was adopted primarily to remedy one of the defects that many people saw in the original Atomic Energy Act of 1946, namely, the prohibition against private ownership of nuclear materials.

With the election of President Dwight D. Eisenhower and a Republican-controlled Congress in 1953, corporate interests received more attention than they had under earlier Democratic-controlled administrations and Congresses. On February 17, 1954, President Eisenhower asked Congress to consider revisions in the Atomic Energy Act of 1946 that would make it easier for the federal government to share information about nuclear energy and to assist private corporations in the development of nuclear facilities than had earlier been the case. In response to this request, Congress passed the Atomic Energy Act of 1954, and the president signed the bill into law on August 30 of that year.

The major purpose of the act was to provide for "the development, use, and control of atomic energy [in such a way] as to promote world peace, improve the general welfare, increase the standard of living, and strengthen free competition in private enterprise." The Atomic Energy Commission was directed to provide information about nuclear science and technology to private industry and to cooperate with private corporations in the development of peacetime applications of nuclear science. The act also instructed the AEC to develop regulations and standards for the design, construction, and operation of nuclear power plants for the protection of human health and the environment, and to establish methods by which these regulations and standards were to be enforced. Detailed instructions about the licensing required for nuclear power plants and other nuclear facilities were provided. Overall, the act significantly expanded the authority and responsibilities of the ACE and the Joint Committee on Atomic Energy. Finally, the AEC was authorized to expand its efforts to work with other nations and international agencies in the development of peacetime applications of nuclear energy.

PRICE-ANDERSON ACT, PUBLIC LAW 85-256 (1957)

Enthusiasm for the development of nuclear power plants in the 1950s was tempered by a number of problems that made private industry reluctant to become involved in such construction. Among the most important of these problems was the legal liability a company would face in case of an accident at a nuclear facility. A study conducted by researchers at the Brookhaven National Laboratory in 1956 (WASH-740) concluded that, in a worst-case scenario, 3,400 people would be killed and 43,000 injured in case of a nuclear accident. In addition, property damage could reach as much as $7 billion. Few, if any, private companies were willing to accept this level of risk in the construction of a nuclear reactor. In addition, the amount of insurance from private companies (even high-risk takers, like Lloyd's of London) available to cover a company's liability in case of a nuclear accident was far too low. The best terms available capped limits at $60 million for liability and an additional $60 million for property damage.

Under these circumstances, the U.S. Congress became convinced that the development of nuclear power in the United States was going to be possible only if the federal government itself assumed all or most of the financial responsibility for accidents that might occur at a facility. As a result, it passed a piece of legislation authored by Senator Clinton Anderson (D-N.Mex.) and Representative Melvin Price (D-Ill.) that absolved private companies from all legal liability for any accidents that might occur at a nuclear power plant. In addition, the Price-Anderson bill allocated $500 million to a fund designed (along with the $60 million available from private insurers) to pay the victims of any such accident. The legislation included an expiration date of 1967, 10 years after its adoption.

As expiration of the original Price-Anderson Act approached, Congress adopted an extension to the legislation in 1966. This extension maintained liability protection for nuclear power plants in essentially the same form as the original act, although it did make some minor adjustments in provisions for liability and payments in case of an accident. The act was amended again in 1975 (for 12 years) and 1988 (for 14 years). As the latest extension was about to expire, Congress took up yet another amendment to the act in 2001. Although the latest extension has not yet been approved, a temporary continuation of Public Law 85-256 was passed by the Congress in 2003.

One of the most significant changes included in the 1975 and 1988 amendments was a shift of primary liability in case of an accident from the federal government to private industry. A fund to cover costs of such an accident was created and paid for by a tax on nuclear power plant owners. Over time, the value of that fund has increased; today it amounts to more than $10 billion, an amount that would be increased to nearly $11 billion if the 2001 amendment passes. For an especially clear and complete review of the history and provisions of the Price-Anderson Act, see "Report on the Price-Anderson Act and its Potential Effects on Eureka County, Nevada" at http://www.yuccamountain.org/price003.htm.

PRIVATE OWNERSHIP OF SPECIAL NUCLEAR MATERIALS ACT, PUBLIC LAW 88-489 (1964)

One of the most serious concerns of legislators in their early debates over the peacetime applications of nuclear energy was the ownership of nuclear materials. With the experience of the first two fission bombs fresh in their minds, public officials worried that these materials might fall into the wrong hands and be used for weapons production. Under both the Atomic Energy

Act of 1946 and the Atomic Energy Act of 1954, therefore, ownership of nuclear materials was restricted to the U.S. government.

As reasonable as this policy may have been from a security standpoint, it proved to be an impediment for industry in the development of nuclear power plants and other facilities. The industry began lobbying almost immediately after World War II, then, for the right to own nuclear materials on their own. Slowly, government officials and legislators were won over to this position and, in 1964, the Congress passed the Private Ownership of Special Nuclear Materials Act, which allowed private companies to purchase and own nuclear materials.

ENERGY REORGANIZATION ACT, PUBLIC LAW 93-438 (1974)

One of the fundamental criticisms that had long been aimed at the Atomic Energy Commission was the inherent conflict in two of its major responsibilities: promoting the use of nuclear power in the United States while adopting and enforcing standards that ensured the safety (and, hence, the cost) of nuclear power plants. In trying to carry out these two somewhat conflicting roles, the AEC was often accused of siding too often with the nuclear industry and too often ignoring the safety problems associated with reactors.

In 1974, the U.S. Congress dealt with this problem by abolishing the AEC and reassigning its responsibilities to two new agencies: the Nuclear Regulatory Commission (NRC) and the Energy Research and Development Administration (ERDA). The former agency was charged with the regulatory functions previously carried out by the AEC. It was assigned the task of regulating the (1) design, construction, and operation of nuclear reactors; (2) research on nuclear materials; and (3) safety and safeguard functions related to nuclear energy. The latter agency was given the task of promoting research and development on nuclear power. ERDA remained in existence for only three years. In 1977, it was abolished as a separate agency and its functions were transferred to the new Department of Energy, created by the Department of Energy Organization Act of 1977. Some critics have suggested that the goal of the 1974 Energy Reorganization Act was never adequately achieved since the NRC continued to have relationships too closely tied to the nuclear power industry to allow it to carry out its regulatory tasks adequately.

URANIUM MILL TAILINGS RADIATION CONTROL ACT, PUBLIC LAW 95-604 (1978)

The Uranium Mill Tailings Radiation Control Act of 1978 (UMTRCA) represented the U.S. Congress's first major attempt to deal with the problem of the dangers posed by radioactivity from materials produced during the fuel cycle. Mill tailings are the materials left over as the result of uranium mining and processing operations. They tend to release only low levels of radiation, but they are produced in such large volumes that, in sum, they may represent a threat to human health and the environment. At one time, mill tailings were used in a variety of construction projects, such as the building of roads.

The UMTRCA directed the Environmental Protection Agency (EPA) to develop methods for reducing the risk posed by mill tailings. Over time, the EPA, usually in cooperation with other agencies, has developed two approaches to the isolation of mill tailings: active methods and passive methods. Examples of active methods of control include the construction of fences and other kinds of barriers around a dump site, posting of signs warning of the dangers of tailings in the area, and adoption of land use regulations that prevent human exposure to wastes. Passive methods of control include the installation of impervious covers on top of a waste dump, preventing the release of radiation to the surrounding atmosphere.

Today, about three dozen mill tailing sites, all, save one, located in the western states, are being monitored by the EPA and other federal agencies. A total of about 200 million metric tons of wastes are stored at these sites. Since the United States now imports the vast majority of uranium used in nuclear reactors and other applications, it is unlikely that additional waste sites will be developed in the future.

LOW LEVEL WASTE POLICY ACT, PUBLIC LAW 96-573 (1980)

Congress's second major attempt to deal with radioactive wastes was the Low Level Waste Policy Act of 1980 (LLWPA). From the late 1940s to the end of the 1970s, low-level wastes were almost always disposed of at the locations at which they were produced: nuclear power plants, research laboratories, medical institutions, weapons production facilities, and industrial plants. In a number of instances, they were disposed of in environmentally dangerous ways, such as dumping into rivers and streams or burying in shallow trenches.

LLWPA established a policy that required states to accept responsibility for the safe disposal of low-level wastes by constructing storage sites either within their own borders or by cooperating with other states to develop "compacts" for that purpose. The act was largely unsuccessful for a number of reasons, one of the most important of which was the reluctance of most states to accept a disposal site within their own borders, although they were usually willing to send their wastes to a site in another state within the compact to which they belonged. In 1985, Congress passed a series of amendments to the LLWPA providing incentives and additional require-ments designed to improve the handling of low-level radioactive wastes. Those amendments were not much more effective than was the original act itself. Nearly all the states did, in fact, join one compact or another, but none of the compacts managed to build a new disposal site. By the end of 2004, only seven sites had ever been licensed to receive low-level wastes by the federal government. Those sites were located at Barnwell, South Car-olina; Beatty, Nevada; Clive, Utah; Hanford, Washington; Maxey Flats, Kentucky; Sheffield, Illinois; and West Valley, New York. Of those sites, only the Barnwell, Clive, and Hanford sites were still accepting low-level wastes for burial.

NUCLEAR WASTE POLICY ACT, PUBLIC LAW 97-425 (1982)

The third in the federal government's trio of acts designed to deal with the nation's nuclear waste problems was the Nuclear Waste Policy Act of 1982 (NWPA), fashioned to deal with high-level wastes. The act charged the De-partment of Energy with responsibility for developing a plan by which the federal government would develop an underground repository for the per-manent storage of high-level wastes. The act also set out a time table for the selection, testing, approval, and opening of a site for the nuclear waste de-pository. According to that timeline, the first wastes were to be delivered to the storage site no later than January 31, 1998.

As with the LLWPA, the NWPA has been largely unsuccessful in solving the problem it was drafted to handle. The selection of a site (Yucca Moun-tain, Nevada) was accomplished, and extensive research on the site has been conducted. But, largely as a result of unexpected environmental problems and the fervent opposition of the state of Nevada and a number of environ-mental groups, progress in construction of the waste repository has gone forward only very slowly. In 2004, development of the site was delayed once again when the Federal Appeals Court for the District of Columbia ruled that the Department of Energy's plans for ensuring the safety of stored wastes for a period of 10,000 years was inadequate. The court ordered DOE to modify its plans to extend the period of time during which the buried wastes could be considered to be safely entombed.

COURT CASES

As is always the case, passing laws is only the first step in establishing the legal framework within which any issue, such as the use of nuclear energy, is resolved. Interpretation of those laws then becomes the responsibility of the court system. Over the past 60 years, a number of important court cases have clarified the legal status of the applications of nuclear energy in the United States. This section summarizes some of the most important cases.

CALVERT CLIFFS' COORDINATING COMMITTEE V. ATOMIC ENERGY COMMISSION, 449 F.2D 1109 (1971)

Background

In the late 1960s, the Baltimore Gas & Electric Company applied for a permit from the Atomic Energy Commission to construct a nuclear power plant in a region known as Calvert Cliffs, on Chesapeake Bay, near Lusby, Maryland. The AEC granted the company a construction permit on July 7, 1969. Before construction could begin, however, a group of concerned citizens who called themselves the Calvert Cliffs Coordinating Committee (CCCC) filed sued to prevent continuation of the project. CCCC claimed that the AEC had not given adequate consideration to the possible environmental effects of the new nuclear power plant on the area around Calvert Cliffs.

Legal Issues

The suit brought by CCCC raised one of the most significant legal issues relating to the use of nuclear energy in the early history of nuclear power plants. Just one year before the suit was filed, the U.S. Congress had passed the National Environmental Policy Act (NEPA) on January 1, 1970. NEPA was arguably the most important single piece of environmental legislation in U.S. history. It provided a broad mandate for maintaining the integrity of the nation's air, water, soil, and other natural resources. By the time CCCC filed suit against the AEC, however, the law had not yet been tested in federal court. Regulatory agencies (among others) were still unclear as to how far NEPA regulations went in restricting the decisions they made that might have environmental consequences.

When the AEC took Baltimore Gas & Electric's application under consideration, it did not ignore possible environmental consequences resulting from the plant's construction. It decided that it could rely on standards other than those provided in NEPA for possible environmental impacts. Specifically, it concluded that Baltimore Gas & Electric was required to show only that it met the standards set by the Federal Water Pollution Control Act of 1948, and not by any further and additional requirements imposed by

NEPA. The AEC argued, fundamentally, that the primary criteria on which permits were to be approved or denied were that a plant met two standards: (1) common defense and security and (2) health and safety of the public. Since the Calvert Cliffs plant appeared to meet these two standards, the AEC granted Baltimore Gas & Electric a permit to build the new nuclear power plant.

Decision

On July 23, 1971, a three-judge panel of the U.S. Court of Appeals for the District of Columbia ruled in favor of the Calvert Cliffs Coordinating Committee. Writing for the panel, Judge J. Skelly Wright made a forceful statement about the reach of the NEPA. He first pointed out that the AEC's decision was a specific violation of NEPA's requirement that environmental impact decisions be made on a case-by-case basis, and that a unique and specific assessment of the environmental effects of the new plant had to be made on a *de novo* basis. He went on to comment on the wide-ranging significance of NEPA for environmental law in the United States, arguing that recent legislation passed by Congress attested to "the commitment of the Government to control, at long last, the destructive engine of material progress." The next step, Judge Wright said, was for the courts to make sure that "important legislative purposes, heralded in the halls of Congress, are not lost or misdirected in the vast halls of the federal bureaucracy." Suits brought by citizens, he said, are the effective tool for ensuring that the objectives of legislation are not "lost or misdirected" within governmental bureacracy.

Impact

The question arose in *Calvert Cliffs' Coordinating Committee v. AEC* whether an agency may, in selecting a rule of general applicability to implement NEPA, defer to a relevant rule prescribed by another agency with environmental expertise. The AEC, in its procedures for implementing NEPA, had provided that a state certification of compliance with water quality standards under the Federal Water Pollution Control Act was sufficient to remove the issue of water quality effects from further consideration in an AEC proceeding for licensing a nuclear power plant. The U.S. Court of Appeals for the District of Columbia held that such automatic deference to another agency's views was inconsistent with AEC's duty under NEPA to consider all environmental factors in its licensing actions. The AEC had based its procedures on two special factors: section 21(b) of the Federal Water Pollution Control Act (added by the Water Quality Improvement Act of 1970), which required the state certification, and congressional statements about the interplay of section 21(b) with NEPA. The appeals court ruled that

NEPA required the AEC to assess water quality effects independently, regardless of a certification of compliance with standards under section 21(b). The court reasoned that by making an "individualized balancing analysis" in each case, the AEC could "ensure that, with possible alterations, the optimally beneficial action is finally taken."

It is not entirely clear whether the AEC or the court of appeals correctly judged the congressional intent concerning the relationship of section 21(b) to NEPA. Legislative clarification of the issue is found in bills since passed by both the House and Senate to amend the Federal Water Pollution Control Act. Those bills carry a provision, supported by the Nixon administration, allowing the AEC and other permit-granting agencies in their NEPA evaluations to rely on state certifications that water quality effects will be acceptable. However, permit-issuing agencies still would be required under NEPA to balance water quality effects along with other factors in making the final permit decision.

The question of whether one agency can defer to another agency's finding of compliance with water quality standards may have limited importance in view of this prompt congressional move to clarify the law. However, it is important to note that, despite the stress in *Calvert Cliffs'* on an "individualized balancing analysis," the opinion does not say that an agency cannot turn to its own general rules to guide all or part of individual decisions. As already pointed out, NEPA requires an agency to balance all competing factors and to consider all reasonable alternatives. It does not dictate that this be done entirely anew in each decision, without the assistance of general rules and past experience. Decision makers are permitted to cut their more complicated decisions down to manageable size. Advance determination of program policy through rulemaking can implement NEPA, at the same time avoiding repetitious reexamination of basic principles in the context of each individual action.

VERMONT YANKEE NUCLEAR POWER CORP. V. NRDC, 435 U.S. 519 (1978)

Background

The Vermont Yankee Nuclear Power Corporation (VYNP) applied for a license from the Atomic Energy Commission for a permit to build a nuclear power plant at Vernon, Vermont, and, in December 1967, received approval from the AEC for the project. Construction began and four years later VYNP applied to the AEC for a license to operate the plant. At this point, the National Resources Defense Council (NRDC), a national environmental organization, filed an objection with the AEC to VYNP's application.

NRDC argued that the company had not adequately considered possible environmental effects related to processing or storage of nuclear fuels at the plant, although it had, in response to AEC regulations, dealt with issues of transporting those fuels. Furthermore, NRDC argued, the AEC should have considered the possibility that the region's energy needs could be met in a more environmentally sensitive way by means other than nuclear power.

The AEC's Atomic Safety and Licensing Board heard NRDC's objections, declined to consider them in making its decision, and, after considering all the evidence presented to it, issued an operating license to YVNP for operation of the Vernon plant. NRDC appealed the Licensing Board's decision to the Court of Appeals for the District of Columbia Circuit, which, in 1976, remanded the AEC's decision to grant a license to VYNP. At that point, VYNP appealed the circuit court's decision to the U.S. Supreme Court, which heard oral arguments on November 28, 1977, and announced its decision on April 3, 1978.

Legal Issues

The case was made somewhat more complex by the fact that, only five months after granting a license to YVNP, the Atomic Energy Commission initiated a revision in its licensing procedure in which it proposed incorporating a more extensive environmental review of the handling of the fuel cycle (which includes processing and storage of nuclear materials) in decisions on the licensing of a new nuclear facility. Recognizing the relevance of this decision to the VYNP decision it had so recently made, the AEC specifically pointed out that any modifications it made in its licensing procedure would not be applicable to that case retrospectively. "The environmental effects of the uranium fuel cycle have been shown to be relatively insignificant," the AEC said, that "it is unnecessary to apply the amendment to applicant's environmental reports submitted prior to its effective date."

The appeals court decision agreed with both of the NRDC's two major contentions: (1) that the environmental consequences of all stages of the nuclear fuel cycle needed to be considered in making a licensing decision, and (2) that ignoring the possibility of alternative sources of energy by the AEC was "capricious and arbitrary."

Decision

In writing a unanimous decision for the Supreme Court, Justice William Rehnquist gave the appeals court a rather severe scolding for having "seriously misread or misapplied this statutory and decisional law cautioning reviewing courts against engrafting their own notions of proper procedures upon agencies entrusted with substantive functions by Congress." That is, the AEC had been entrusted by Congress with establishing its own regulatory system for deciding on licenses for nuclear power plants, and courts had no latitude in deciding whether or not they thought those regulations were adequate in making such decisions. The appeals court was simply wrong, Rehnquist said, for having "unjustifiably intruded into the administrative process" and erred in "depart[ing] from the very basic tenet of administrative law that agencies should be free to fashion their own rules of procedure." The AEC knew best what procedure to follow in making decisions on nuclear power plant licenses, and as long as those procedures did not violate the letter or spirit of the enabling legislation, the courts had no right to invalidate or modify those procedures. The Court reversed the appeals court decision, and the AEC's license for the Vermont Yankee plant was affirmed.

Impact

This decision had relatively little effect on the process by which nuclear power plants are licensed. By the time the decision had been announced, the AEC had adopted a modification of its licensing procedure that incorporated an effort to assign a numerical value to the possible environmental effects of various steps in the nuclear fuel cycle. This modification, currently known as Table S.3, is contained in the Nuclear Regulatory Agency's *Generic Environmental Impact Statement for License Renewal of Nuclear Plants* (NUREG-1437, Vol. 1). Thus, the case that NRDC originally made for obtaining a more comprehensive analysis of the environmental effects of the nuclear fuel cycle was, in fact, eventually achieved, although it had no effect on the specific licensing decision at Vermont Yankee.

More significant, perhaps, were two other points about the Court's decision. First, Justice Rehnquist used the case to speak very forcefully about "activist" decisions by lower court judges, decisions in which those judges went beyond their assigned role of judicial review and superimposed their own opinions and beliefs on regulatory agencies. Second, Rehnquist reiterated the frequently expressed view of Court authors that the risk of environmental hazards from the use of nuclear fuels was so small that it could legitimately be ignored in making decisions about the licensing and operation of nuclear power plants.

DUKE POWER CO. V. CAROLINA ENVIRONMENTAL STUDY GROUP, 438 U.S. 59 (1978)

Background

In 1957, the U.S. Congress passed the Price-Anderson Act (P.L. 85-256), limiting the liability of private industries that owned and operated nuclear power plants to the $60 million for which they could obtain private insurance. The act then established a public fund in the amount of $500 million to supplement the costs of any accident that might exceed the amount available from private insurance. Congress passed the Price-Anderson Act because it had become convinced that no private company would undertake the construction of a nuclear facility if it could not obtain insurance adequate to protect itself in case of an accident. And no private insurance company had been willing to offer more than $50 million in coverage. If nuclear power was to have a future in the United States, then, Congress decided, the federal government would have to assume a substantial portion of the costs needed to provide coverage in case of an accident.

Legal Issues

In 1973, an environmental organization (the Carolina Environmental Study Group), a labor union (the Catawba Central Labor Union), and a group of 40 individuals living near the proposed site of a new nuclear power plant in North Carolina sued Duke Power Company, owner of the proposed plant, and the Nuclear Regulatory Commission, claiming that the Price-Anderson Act was unconstitutional and asking that construction be halted on the plant. The appellees' argument was that the amount of liability coverage provided by the Price-Anderson Act was insufficient to pay all claims that might arise as the result of an accident at a nuclear facility.

The district court to which the case was assigned agreed with the plaintives. The court concluded that the Price-Anderson Act contravened the Due Process Clause of the Fifth Amendment of the U.S. Constitution because "[t]he amount of recovery is not rationally related to the potential losses"; because "[t]he Act tends to encourage irresponsibility in matters of safety and environmental protection. . ."; and finally because "[t]here is no quid pro quo" for the liability limitations. Duke Power appealed the district court's decision to the U.S. Supreme Court, which heard the case argued on March 20, 1978, and announced its decision on June 26, 1978.

Decision

Justice Warren Burger wrote the opinion for the Court, a decision that was signed or concurred in by all members of the Court. In his opinion, Burger found that the district court was in error and that the Price-Anderson Act was not unconstitutional. Burger emphasized three points on which the Court's decision was based. First, the Price-Anderson Act was, in fact, "rationally related" to Congress's intention of stimulating the development of nuclear power by private industry. Second, the chances of an accident at a nuclear power plant are so remote that a $560 million liability limit is reasonable and does not violate the Due Process rights of the original plaintives. Finally, Congress expressed its intent in the Price-Anderson Act to "take whatever action is deemed necessary and appropriate to protect the public from the consequences of" a disaster of such proportions that the $560 million liability limit would be inadequate. Therefore, Congress had guaranteed a level of protection that was sufficient for any nuclear accident of any size.

Impact

The significance of the *Duke* decision for the future of nuclear power in the United States can hardly be overestimated. Had the Court agreed with the Carolina Environmental Study Group and its partners in this case, the fundamental principle underlying the Price-Anderson Act would have been negated, and the federal government would have had to find a new way of protecting industry from legal liabilities with sufficient strength to encourage industry to continue with its development of nuclear power plants. As it happens, the Court was correct in believing that the risk of nuclear accidents at a power plant were "so remote" that the actual dollar amount of liability involved was probably not very important. In more than five decades, no member of the general public has ever been injured in a nuclear accident in the United States and, as a consequence, industry has never had to face the problem of legal liability for its actions.

METROPOLITAN EDISON V. PEOPLE V. NUCLEAR ENERGY, 460 U.S. 766 (1983)

Background

At 4:00 A.M., on March 28, 1979, a failure in the condensing system in one of the steam generators at the Three Mile Island Unit 2 (TMI-2) nuclear power plant near Middletown, Pennsylvania, failed, setting off a series of reactions that resulted in the worst nuclear power plant disaster in the United States. Although no deaths were attributed to the accident, critics of the nuclear power industry have almost unanimously used the Three Mile Island accident as evidence of the risks posed by the technology. In response to the accident at TMI-2, the Nuclear Regulatory Commission ordered the plant's owner, Metropolitan Edison, to close down the reactor's companion plant, Unit 1 (TMI-1). TMI-1 had, coincidentally, been closed down for refueling on the day TMI-2 experienced its malfunctions.

On August 9, 1979, the Nuclear Regulatory Commission announced hearings on its intention to permit Metropolitan Edison to restart Unit 1 at Three Mile Island and invited comments from the general public on this decision. It indicated that it had not decided whether or not to consider the issues of psychological harm and/or indirect damage to the public in its decision about reopening TMI-1. But it offered to entertain briefs on these two points.

In response to this invitation, a group of residents living in the area near Three Mile Island, organized under the name Respondent People Against Nuclear Energy (PANE), submitted a brief detailing psychological and other indirect effects that the TMI-2 accident had had on the community. It offered this brief in opposition to the possible restarting of TMI-1.

By the time restarting hearings were actually held, the NRC had decided not to consider psychological or other indirect factors and declined to include PANE's brief in its considerations. In response to this decision, PANE filed suit with the U.S. Court of Appeals for the District of Columbia District, claiming that the NRC's actions had violated both the Atomic Energy Act of 1954 (AEA) and the National Environmental Protection Act of 1970 (NEPA). The court of appeals decided that the NRC was not bound by any conditions of the AEA in this case, but that NEPA did impose certain requirements on the commission, specifically that it "evaluate the potential psychological health effects of operating" the plant, especially those that may have arisen as a result of the TMI-2 accident, long after the original environmental impact statement had been prepared and approved. The NRC, joined by Metropolitan Edison, appealed the appeals court's decision to the U.S. Supreme Court, which heard oral arguments on the case on March 1, 1983.

Legal Issues

The fundamental question facing both courts was what Congress had intended when it referred to the "environmental impact" of some action. In a core segment of NEPA, for example, Congress requires all federal agencies to "include in every recommendation or report on proposals for legislation and other major Federal actions significantly affecting the quality of the human environment, a detailed statement by the responsible official on (i) the environmental impact of the proposed action, [and] (ii) any adverse environmental effects which cannot be avoided should the proposal be implemented" [42 U.S.C. 4332(C)]. The question is to what extent an "environmental impact" may extend. Does it include only physical impacts, such as the production of chemical changes in water and air? Or does it extend to the cultivation of neuroses, psychoses, and/or other mental disorders?

Decision

The appeals court accepted the broader interpretation of NEPA. In its decision on this case, the court instructed the NRC to develop a new "supplemental (environmental impact statement) which considers not only the effects on psychological health but also effects on the well-being of the communities surrounding Three Mile Island." Justice William Rehnquist, writing for a unanimous Supreme Court, rejected that view. A review of the history of NEPA, he said, clearly shows that Congress had in mind the physical environment when they were writing the law. Rehnquist quoted Senator Henry Jackson (D-Wash.), for example, who said that the purpose of NEPA was to ensure that "we will not intentionally initiate actions which do irreparable damage to the *air, land and water* which support life on earth" (emphasis added). At no point in Congress's deliberation was there any intent expressed to go beyond this level of environmental impact, to the kinds of mental and emotional harm suggested by the PANE argument. Based on this analysis, the Supreme Court reversed the appeals court decision and permitted the NRC to continue with its restarting of the TMI-1 plant.

Impact

The primary consequence of the Court's action in this case was to clarify the responsibility that the NRC (and other federal agencies) had in meeting the "public safety and protection" provisions of the Atomic Energy Act of 1954 and the "environmental protection" provisions of the National Environmental Protection Act of 1970. Its responsibilities extended exclusively to those areas in which an impact can be measured by some objective means, such as physical, chemical, geological, or biological characteristics, and not to somewhat more subjective properties such as emotional state or community cohesiveness.

SILKWOOD V. KERR-MCGEE CORPORATION, 464 U.S. 615 (1984)

Background

Karen Silkwood was a laboratory technician at a nuclear production facility operated by the Kerr-McGee Corporation's Cimarron plant near Crescent, Oklahoma. The primary activity at the plant was fabrication of plutonium fuel elements used in the reactor core of nuclear power plants. In November 1974, Silkwood discovered that she had become contaminated with radioactive particles, apparently as the result of her work in the grinding and polishing of plutonium metal. Follow-up studies by plant health monitors showed that Silkwood's contamination was far more extensive than at first imagined, with both her urine and fecal samples showing high levels of radioactivity. Examination of Silkwood's apartment also showed high levels of contamination of her clothing, appliances, and other materials present. Even her roommate, a fellow worker at the plant, was found to be contaminated with plutonium.

The Silkwood case was far more complex than one might imagine from reading the legal documents related to the Supreme Court case discussed here. She was an active member of the union at the Cimarron plant and served on the union's bargaining committee. She had expressed some concern about safety practices at the plant and had reportedly offered to provide evidence of unsafe activities at the plant to union officers and a reporter from the *New York Times* at about the time she was found to be contaminated. That evidence never surfaced, however, and Silkwood was killed in a one-car accident on November 13, 1974, presumably en route to a meeting with the union official and reporter. One major book and an important motion picture were later released pursuing the question as to the relationship of Silkwood's death, her contamination with plutonium, and her union activities at the plant. None of these issues are part of the case reported here, however.

Legal Issues

This court case arose when Silkwood's father, Bill Silkwood, sued the Kerr-McGee company, arguing that her death was the result of unsafe practices at the plant. The jury in the original trial agreed with Silkwood, and awarded him $500,000 for personal injuries, $5,000 for property damage, and $10 million in punitive damages. Kerr-McGee then appealed to the Tenth Circuit Court of Appeals, which vacated the trial court's decision. The key issue at hand, and the basis for Kerr-McGee's appeal, was that Congress had made the Atomic Energy Commission, and later its successor, the Nuclear Regulatory Commission, the sole agency in the United States responsible for the regulation of radiation hazards. The appeals court concluded that "any state action that competes substantially with the AEC (NRC) in its regulation of radiation hazards associated with plants handling nuclear material" was impermissible. It eventually affirmed the $5,000 property damage to Silkwood; reversed the $500,000 damage for personal injuries, on the basis that personal injuries were covered by Oklahoma worker's compensation law; and reversed the $10 million punitive damage, on the basis that the award violated the exclusive responsibility of the AEC (NRC) to administer laws and regulations dealing with radiation. Bill Silkwood then appealed the appeals court decision to the U.S. Supreme Court, which heard arguments on October 4, 1983, and announced its decision on January 11, 1984. The fundamental question considered by the Court was whether federal law did, in fact, supersede state action in cases involving radiation hazard issues.

Decision

The Court decided that the Appeals Court had erred in its decision and restored the original trial court's decision in the case. Writing for a 5 to 4 majority, Justice Byron White explained that the awarding of punitive damages by the Oklahoma court did not interfere with the rights and responsibilities of the NRC and "is not physically impossible, nor does exposure to punitive damages frustrate any purpose of the federal remedial scheme." The fundamental principle involved, according to Justice White, is that, in its enabling legislation for the AEC (and NRC), Congress understood that it might be possible for an individual harmed by a nuclear hazard to be subject to both federal penalties and the kind of legal remedies requested by Bill Silkwood in this case. Indeed, White pointed out, it "is difficult to believe" that Congress would simply remain silent and, by its inaction, remove all legal remedies to a person injured in a nuclear accident. Finally, the basic intention of Congress has, all along, been the safety and protection of the general public, and the trial court's decision in no way conflicts with that intention.

Impact

The most important consequence of the *Silkwood* decision was that it clarified the legal options available to a person (or persons) injured as the result of a nuclear hazard in the United States. On the one hand, it is clearly the federal government's responsibility in general, and that of the NRC in particular, to regulate and oversee the operation of a nuclear facility in such a way as to ensure "the safety and protection of the general public." On the other hand, the assignment of this function to the NRC does not preclude the possibility that someone injured in a nuclear accident may also seek a variety of legal remedies, such as compensation for personal and property damage and punitive penalties.

Beyond the field of nuclear hazards, *Silkwood* has formed an important pillar in Supreme Court philosophy regarding the question of the circumstances under which federal law precludes legal action by individuals and states and the conditions under which such actions may be permitted and/or appropriate. In more than four dozen major cases decided since *Silkwood*, for example, the Court has referred to that decision as forming at least part of the precedent for its decisions.

ALLEN V. UNITED STATES, 816 F.2D 1417 (1987)

Background

Under provisions of the Atomic Energy Act of 1946, the Atomic Energy Commission was authorized to carry out tests of nuclear weapons. In 1951, the AEC selected a site in Nevada at which such tests were to be conducted. The site, covering an area near Las Vegas that was larger than the state of Rhode Island, became known as the Nevada Test Site (NTS). Between 1951 and 1962, a total of 105 atmospheric tests of nuclear weapons were conducted at that site. By the mid 1960s, the NTS had become one of the most radioactive places on Earth.

The radiation produced during weapons testing did not remain within bounds of the NTS, however. Prevailing westerly winds carried high levels of radiation to the east, across southern Utah and northern Arizona. The first indication of the possible effects of this spreading radiation was widespread illness and death among cattle and sheep in Utah. By the 1970s, however, health effects were also being observed among humans in the area. In southern Utah, for example, the incidence of various types of cancer increased significantly, with a much higher death rate from those diseases being recorded.

In August 1979, a group of 1,192 individuals brought suit in the U.S. District Court for Utah on behalf of themselves and their relatives, alleging that the diseases and deaths they had experienced were the result of improper practices on the part of the U.S. government and its employees during the bomb tests of 1951–62. The suit was brought under the Federal Tort Claims Act (28 U.S.C. 1346(b), 2401(b), 2671–80), which allows private citizens to sue the government (which is normally immune from lawsuits) when its employees are negligent in the conduct of their duties. The district court decided to select 24 of the 1,192 cases as "bellwether" cases on which decisions on the remaining cases might be based. The case was eventually named after the first of the 24 litigants (alphabetically), Irene H. Allen. Discovery in preparation for the trial took more than two years, and the trial itself did not begin until September 20, 1982. The trial lasted 13 weeks, producing a transcript of more than 7,000 pages in length, with exhibits covering an additional 54,000 pages. The court deliberated for 17 months before issuing a 225-page opinion.

Legal Issues

The court was faced with two major issues in this case. The first involved the matter of the government's liability for any injury that plaintiffs may have sustained as a result of the bomb tests it conducted at NTS. In general, governments tend to take the position that they are not legally liable for damage that results from their actions, except within very special circumstances, such as those defined by the Federal Tort Claims Act (FTCA). The court had to decide whether this case fell within the narrow boundaries under which the federal government could be sued under the FTCA. Second, the court had to decide whether the injuries suffered by the plaintiffs (such as an increased incidence of cancer and death) could clearly be associated with bomb testing and not with any number of other factors.

Decision

The district court agreed broadly with the case presented by the 24 plaintiffs. Judge Bruce Jenkins accepted the argument that an agency or agencies of the federal government, and not just its on-site employees, were negligent in not providing the proper degree of safety to protect the public from harm as a result of radioactive fallout. Jenkins also agreed that a legitimate connection could be made between the diseases experienced by at least some of the plaintiffs and radioactive fallout produced by bomb testing. He awarded judgments to 10 of the 24 plaintiffs on these grounds, but denied a similar judgment to the remaining 14 plaintiffs.

The federal government appealed Jenkins's ruling to the U. S. Court of Appeals for the Tenth Circuit, which overturned the district court's decision. The appeals court had the additional benefit in reaching its decision of a Supreme Court opinion on governmental liability (*United States v. S.A. Empresa de Viacao Aerea Rio Grandense [Varig Airlines]*, 467 U.S. 797) issued after the district court's hearing of the case. Leaning heavily on the Court's reasoning in this case, the appeals court decided that the federal government could not be held liable for inadequate or improper safety practices conducted by its employees at the NTS range, and that plaintiff's injuries, as far as they could be connected with those practices, were not the result of federal policy. Given its decision that the federal government has "broad sovereign immunity," the complex line of reasoning that had allowed Judge Jenkins to allow judgment to some plaintiffs and not to others (because they had proved or not proved the relationship between disease and fallout) became moot. Since they couldn't sue the government, there was no point in the plaintiffs' trying to show how their conditions were related to governmental action.

Impact

With regard to the question of governmental liability for damage to individual citizens, this case had relatively little impact on future cases. As Justice Monroe G. McKay wrote in a concurring opinion of the appeals court, "It undoubtedly will come as a surprise to many that two hundred years after we threw out King George III, the rule that 'the king can do no wrong' still prevails at the federal level in all but the most trivial of matters." But, he went on to say, that's the way it is, and courts have continued to give the government wide latitude of action in tort cases brought against it.

Perhaps of greatest interest in the case, however, was Judge Jenkins's careful analysis of the way in which courts can attack the question of how certain types of damage (such as illness and death) can be associated with certain types of causes (such as nuclear weapons testing). Jenkins established the principle in this case that "a population exposed to a certain dose of radiation will show a greater incidence of cancer than that same population would have shown in the absence of the added radiation." All the plaintiffs had to do, then, was to show that the condition from which they suffered might reasonably result from an increased exposure to radiation. At that point, Jenkins said, the burden of proof fell to the government to show that their acts could not be responsible for a plaintiff's medical problems. While this view has not become a predominant theme in American jurisprudence, it has, nonetheless, provided a somewhat different philosophy from which to analyze the cause-and-effect association that is problematic in many cases.

NUCLEAR ENERGY INSTITUTE, INC. V. ENVIRONMENTAL PROTECTION AGENCY, U.S. COURT OF APPEALS FOR THE DISTRICT OF COLUMBIA CIRCUIT, NO. 01-1258 (2004)

Background

In 1982, the U.S. Congress passed the Nuclear Waste Policy Act (P.L. 97-425) establishing national policy for the treatment and disposal of high-level radioactive wastes and setting a timeline by which a permanent repository for such wastes in the United States was to be identified, researched, and put into use. In 1987, recognizing that the conditions of the original act were resulting in a process that was far too lengthy and expensive, Congress narrowed the selection of sites from five (as originally proposed in the NWPA) and then three (as had been determined by 1985) to one: Yucca Mountain, in Clark County, Nevada. In the Nuclear Waste Policy Amendments Act of 1987 (Public Law 100-203), Congress directed that the Nuclear Regulatory Commission, Environmental Protection Agency, Department of Energy, and other relevant agencies focus henceforth on Yucca Mountain as the only candidate for a high-level radioactive waste disposal site in the United States.

In the two decades following that decision, the controversy over using Yucca Mountain as a waste disposal site continued, with vigorous opposition to Congress's decision arising largely from a number of environmental groups, the State of Nevada, and various local governmental units (such as the City of Las Vegas and Clark County). One series of lawsuits eventually found its way to the U.S. Court of Appeals for the District of Columbia Circuit. A total of 13 separate cases were consolidated into a set of four cases, for which oral arguments were heard on January 14, 2004.

Legal Issues

The cases heard by the court presented four distinct types of challenge to the selection of Yucca Mountain as a waste disposal site. First, the Nuclear Energy Institute, Inc., challenged the EPA's groundwater standards as being both unnecessary and illegal. Second, the state of Nevada, Clark County, and the city of Las Vegas challenged the Nuclear Waste Policy Amendments Act of 1987 as being unconstitutional because they required that a single county in a single state be responsible for the storage of all the nation's high-level nuclear wastes. Third, the same three governmental units challenged the site-suitability standards selected by the Department of Energy and the DOE's final environmental impact statement. Fourth, a number of environmental groups and the state of Nevada challenged the EPA's radiation-protection standards as being insufficient to guarantee the protection of public health and safety.

Decision

The court rejected the first three of the four challenges outlined above. It said that (1) the environmental standards established by the EPA, with one exception to be noted below, are neither unlawful nor arbitrary; (2) Congress was entirely within its authority to designate a single specific site for nuclear waste storage; and (3) the series of steps that led to the selection of the Yucca Mountain site by Congress and the president are entirely legal and not subject to review.

The only point on which a challenge was upheld related to the EPA's selection of a 10,000-year compliance period for the storage site. The 10,000-year compliance period refers to a decision by the EPA that any disposal and storage method developed for the Yucca Mountain site must guarantee that the general public would be protected from harmful radiation for a period of at least 10,000 years. The choice of a 10,000-year period was based by the EPA on a 1995 study conducted by a committee of the National Academy of Sciences (NAS), mandated by Congress in the 1992 Energy Policy Act. In its 1995 report, *Technical Bases for Yucca Mountain Standards*, the NAS committee pointed out that the risks posed by high-level nuclear wastes is likely to extend from tens to hundreds of thousands of years after disposal or even further into the future. In addition, the committee said, it should be possible to construct a safe site in the proper geological terrain that would provide protection to the public for a million years or so.

In spite of these recommendations, the EPA chose to set 10,000 years as the period of time over which safe storage would be guaranteed. It pointed out that requiring secure storage for a longer period of time, although scientifically feasible, would not be practical for regulatory decisionmaking. The Department of Energy and the Nuclear Regulatory Commission agreed with EPA's decision, although the state of Nevada and various environmental groups objected, and argued that the period was too short.

The appeals court agreed with the state of Nevada and the environmental groups. It said in its decision of July 9, 2004, that the EPA's choice of a 10,000-year compliance period was not based upon or consistent with the recommendations made by the NAS committee, which was appointed specifically for the purpose of making such recommendations. It concluded that

> *because EPA's chosen compliance period sharply differs from NAS's findings and recommendations, it represents an unreasonable construction of section 801(a) of the Energy Policy Act [requiring EPA to set standards consistent with an NAS study] . . . [and] [w]e will thus vacate part 197 to the extent that it requires DOE to show compliance for only 10,000 years following disposal. . . . On remand, EPA must either issue a revised standard that is "based upon and consistent with" NAS's findings and recommendations or return to Congress and seek legislative authority to deviate from the NAS Report.*

Impact

The court's ruling was received with mixed reactions from both sides of the Yucca Mountain controversy. On the one hand, Secretary of Energy Spencer Abraham released a statement in which he said that he was "pleased" with the court's decision. That decision, Abraham said, confirmed that the fundamental plan for storing wastes at Yucca Mountain was constitutional and scientifically sound. He acknowledged that additional work would be necessary to meet the court's higher standard for storage of radioactive materials. By contrast, Brian Sandoval, Nevada's attorney general, headlined his press release of July 9, 2004, "Sound Science Trumps Yucca Mountain." He said that the Department of Energy would not be able to meet the kind of standard demanded by the court, one in which safe storage for hundreds of thousands of years or more was necessary. Opponents of the Yucca Mountain site were convinced that the court's decision would either delay or bring an end to the project.

CHAPTER 3

CHRONOLOGY

This chapter presents a chronology of major events in the history of nuclear energy, including scientific, technological, economic, and political happenings related to both the military and peacetime applications of nuclear energy.

circa 400 B.C.

- Greek natural philosopher Democritus (about 370–460 B.C.E.) argues that matter consists of tiny, indivisible particles that he calls *atomos* ("indivisible").

1803

- English chemist and physicist John Dalton outlines the major assumptions of the modern atomic theory.

1895

- William Roentgen, a German physicist, discovers X-rays, a form of electromagnetic radiation with a shorter wavelength (and, therefore, more energetic wavelength) than light waves.

1896

- French physicist Antoine Henri Becquerel accidentally discovers radioactivity while studying the use of Roentgen's X-rays.

1897

- British physicist J. J. Thomson discovers the electron, thereby providing the first experimental evidence that atoms are not indivisible, but they consist of at least two distinct parts.
- British physicist Lord Ernest Rutherford discovers that the radiation emitted by radioactive materials consist of at least two distinct types, which he names alpha and beta rays.

1898

- French physicists and chemists Marie and Pierre Curie give the name radioactivity to the process by which certain elements spontaneously release radiation and break down into simpler elements. During their research on radiation, the Curies also discover two new elements, radium and polonium, both of which are radioactive.

1900

- French physicist Paul Villard discovers gamma rays.

1905

- Austrian physicist Albert Einstein develops a general theory of relativity, one aspect of which reveals the inherent relationship between matter and energy. The mathematical formula that he derives for this relationship is $E = mc^2$.

1919

- Lord Ernest Rutherford conducts a series of experiments that results in the first artificial transmutation of elements, that is, the conversion of one element into a different element. In his work, Rutherford bombards nitrogen with alpha rays and discovers that oxygen is formed in the process.

1920

- In the period between 1911 and 1914, Lord Ernest Rutherford and his assistant, English physicist Frederick Soddy, discover the proton, for which no name is agreed upon until Rutherford suggests the term in 1920.

1923

- Hungarian chemist Georg von Hevesy suggests that radioactive isotopes can be used as tracers, materials whose presence in and movement through a system can be observed because of the radiation they emit.

1932

- British physicist James Chadwick discovers the neutron, completing the set of three fundamental particles (proton, neutron, and electron) of which atoms are made.

1936

- American physicist John H. Lawrence, at the University of California at San Francisco, uses radioactive phosphorus, phosphorus-32, to treat leukemia, the first documented case in which a radioactive isotope is used to treat a disease.

1938

- German physicists Otto Hahn and Fritz Strassman discover the process of nuclear fission when they bombard uranium with neutrons and find that barium, krypton, and other smaller nuclei are formed. Reluctant to accept the apparent results of their experiment, they ask their colleague Lise Meitner to develop a theoretical explanation of their experiment. Meitner and her nephew Otto Frisch confirm that nuclear fission has, in fact, occurred in the experiment.
- American chemist Glenn Seaborg and Italian physicist Emilio Segré discover the radioactive isotope technetium-99m, now the most widely used of all radioisotopes in the field of medicine.

1939

- *March 17:* Italian physicist Enrico Fermi presents an address to the U.S. Navy's Technical Division about the possibility of nuclear weapons based on the fission reaction. Navy personnel in general appear to be little interested in Fermi's talk.
- *November 11:* Albert Einstein and Leo Szilard deliver a letter to President Franklin D. Roosevelt outlining the military potential of nuclear fission weapons. The letter apparently has little impact on the president or his advisers, and no action is taken until word begins to reach Washington that German scientists may already be working on such a weapon.

1940

- A research team led by Glenn Seaborg, at the University of California at Berkeley, discovers the transuranium element plutonium. One isotope of the element, plutonium-239, turns out to be one of the few isotopes that is fissionable and, hence, suitable for the construction of nuclear weapons. Plutonium was the fissionable fuel used in the production of Fat Man, the bomb later dropped on Nagasaki.

1941

- *October 9:* President Franklin D. Roosevelt authorizes the initiation of a secret research project for the manufacture of nuclear weapons, a project given the code name of Manhattan Engineering District, later to be known as the Manhattan Project.

1942

- *December 2:* The world's first controlled nuclear chain reaction is achieved by a research team at the Chicago Metallurgical Laboratory working in a converted squash court under the university's football stadium, Stagg Field.
- The first successful use of iodine-131 for the treatment of hyperthyroidism is reported by a research team from the Thyroid Clinic of the Massachusetts General Hospital, the Boston George Eastman Research Laboratory at the Massachusetts Institute of Technology, and the Boston Thorndike Laboratory of Boston City Hospital. In the same year, a seminal paper on the "The Use of Radioactive Tracers in Biology and Medicine," by American physicist Joseph G. Hamilton, appears in the journal *Radiology*.

1945

- *July 16:* The first successful test of a fission weapon is conducted at Alamogordo, New Mexico, under the code name Trinity.
- *August 6:* The world's first nuclear weapon, a fission bomb containing uranium-235, is dropped on Hiroshima, Japan.
- *August 9:* The world's second nuclear weapon, a fission bomb containing plutonium-239, is dropped on Nagasaki, Japan.
- *December:* A group of scientists working on the Manhattan Project and concerned about the hazards posed by the new technology found *The Bulletin of the Atomic Scientists*, a publication that is destined to become one of the most highly respected voices analyzing the risks posed by nuclear weapons and nuclear power plants.

1946

- *May 26:* The U.S. Army Air Force awards a contract to the Fairchild Engine and Airplane Corporation for the development of a nuclear-powered airplane. The U.S. Congress had authorized the creation of the Nuclear Energy Propulsion Aircraft project earlier that same year.
- *August 1:* President Harry S. Truman signs the Atomic Energy Act of 1946.
- *December 25:* The first full-scale nuclear reactor built outside the United States begins operation at the Kurchatov Institute in Moscow. The reactor is still in operation.

1947

- *January 1:* The U.S. nuclear energy program is transferred from control by the U.S. military under the Manhattan Engineering District to the newly created Atomic Energy Commission.

- The Atomic Energy Commission establishes the Reactor Safeguards Committee (RSC) in order to monitor the safety of nuclear power plants in the United States.

1949

- *March 1:* The Atomic Energy Commission announces the selection of Arco, Idaho, as the location for a National Reactor Testing Station, at which research on the construction and operation of nuclear power plants is to be carried out.
- *August 29:* The Soviet Union detonates its first nuclear weapon, a fission bomb, at its testing site at Semipalatinsk, Kazakhstan.

1950

- *January 31:* U.S. president Truman announces that he has authorized American scientists to begin work on the construction of a fusion ("hydrogen") bomb.
- The Atomic Energy Commission creates an Industrial Committee on Reactor Location Problems (ICRLP) to evaluate hazards associated with the operation of nuclear power production facilities.

1951

- *December 20:* The Experimental Breeder Reactor 1 (EBR-1) at the National Reactor Testing Site in Arco, Idaho, produces the first electrical power obtained from nuclear fission, sufficient power to operate four household-size light bulbs.

1952

- *June 14:* Construction is started on the world's first nuclear submarine, the *Nautilus,* at the Electric Boat Division of General Dynamics Corporation in Groton, Connecticut.
- *October 3:* Great Britain conducts its first test of a fission bomb at the Monte Bello Islands, off Australia.
- *November 1:* The United States conducts the first test of a fusion bomb on the small island of Eniwetok in the Marshall Islands.
- *December 12:* One of the world's first major nuclear power accidents takes place at Canada's NRX nuclear reactor, located at Chalk River, Ontario. The reactor was used primarily for the production of plutonium for the U.S. military. During the accident, the reactor core experienced a partial meltdown, during which radiation was released to the surrounding environment. No human deaths or illnesses were attributed to the accident.

1953

- *July:* The Atomic Energy Commission combines the Reactor Safeguard Committee and the Industrial Committee on Reactor Location Problems under the new name of the Advisory Committee on Reactor Safeguards.
- *August 12:* The Soviet Union conducts its first test of a fusion bomb at its Semipalatinsk Test Site.
- *December 8:* President Dwight D. Eisenhower announces an Atoms for Peace plan in a speech before the United Nations.

1954

- *January 21:* The nuclear submarine *Nautilus* is launched.
- *July 1:* The world's first commercial nuclear power plant begins operation at Obinsk, Russia, in the Soviet Union, about 60 mi (100 km) south of Moscow.
- *August 30:* President Eisenhower signs the Atomic Energy Act of 1954.
- *September 6:* Ground is broken for construction of the Shippingport Nuclear Power Station, a joint endeavor of the federal government and the Dusquene Light Company of Pittsburgh, Pennsylvania.

1955

- *January 10:* The Atomic Energy Commission announces the creation of the Power Demonstration Reactor Program, a cooperative program between the federal government and private industry for the development of experimental nuclear reactors.
- *July 17:* The town of Arco, Idaho, becomes the first community in the world to have its electrical needs met entirely by nuclear energy. The electricity is provided by an experimental boiling water reactor, BORAX III.
- The Atomic Energy Commission asks the National Academy of Sciences (NAS) to undertake a study of the problem of nuclear waste disposal. Two years later, a NAS committee, the Committee on Waste Disposal, issues a report recommending that nuclear wastes be buried deep within geological formations known as salt domes.

1956

- *May 4:* The Atomic Energy Commission authorizes construction of the first privately owned nuclear power plants. The plants constructed as a result of this agreement are the Indian Point Nuclear Power Plant, in Buchanan, New York, built by Consolidated Edison Company of New

York; and the Dresden 1 Nuclear Power Station, in Grundy County, Illinois, built by Commonwealth Edison.

1957

- *March:* Researchers at Brookhaven National Laboratory produce a report entitled "Theoretical Possibilities and Consequences of Major Accidents in Large Nuclear Plants," the so-called WASH-740 report. The report estimates that a nuclear power plant accident could result in about 3,000 deaths, 43,000 injuries, and property damage of about $7 billion.
- *May 15:* Great Britain conducts its first test of a fusion bomb.
- *July 12:* The first experimental civilian nuclear power reactor, the Sodium Reactor Experiment located at Santa Susana, California, begins generating power. The reactor remains in service until 1966.
- *August 3:* The Vallecitos Boiling Water Reactor, located near Pleasanton, California, begins operation. The reactor is issued License No. 1 by the Atomic Energy Commission. It is connected to the utility grid on October 19, 1957, and operates until December 9, 1963. The primary purpose of building the Vallecitos plant was to gain experience in preparation for the construction of the first major privately funded and constructed nuclear power plant to be built by Commonwealth Edison at Dresden, Illinois.
- *September 2:* The Price-Anderson Act is signed by President Eisenhower. The act provides financial protection to nuclear power companies in case of a major accident. The act is designated as Public Law 85-256 and set to expire in 1987. A year after its expiration date, it is reinstated for an additional 15-year period.
- *October 1:* The United Nations establishes the International Atomic Energy Agency (IAEA), headquartered in Vienna, Austria, for the purpose of promoting the peaceful applications of nuclear energy and preventing the spread of nuclear weapons throughout the world.
- *October 10:* A fire breaks out at the Windscale Nuclear Power Plant north of Liverpool, England, resulting in the release of radiation to the surrounding environment. The British National Radiological Protection Board later estimated that the accident resulted in 32 deaths and at least 260 cases of cancer.
- *December 2:* The world's first large-scale commercial nuclear power plant, the Shippingport Nuclear Power Station, in Shippingport, Pennsylvania, begins operation. It takes three weeks for the plant to reach maximum operational capacity. The plant is later taken out of operation and decommissioned in 1989.

Nuclear Power

1958

- *May 22:* Construction begins on the world's first nuclear-powered surface ship, the NS *Savannah*, in Camden, New Jersey. She is later launched on March 23, 1962, removed from active service in 1972, mothballed in 1985, and moved to the James River Merchant Marine Reserve Fleet in 1999.
- *May 23:* A second accident occurs at the nuclear reactor located at Chalk River, Ontario. A control rod catches fire after being removed from the reactor, releasing radioactivity to the interior of the plant.

1959

- *October 15:* The Dresden 1 Nuclear Power Station begins operation when its nuclear reactor achieves a self-sustaining nuclear reaction. The plant, built to serve the city of Chicago and surrounding areas, is removed from service in 1978.

1960

- *February 13:* France tests its first fission bomb at the Reggane testing station in the Sahara Desert.

1961

- *January 3:* An incident at the National Reactor Testing Station results in the death of three technicians, the first nuclear accident involving a fatality in U.S. history.
- *March 23:* The NS *Savannah* is launched. The *Savannah* is the only U.S. nuclear-powered cargo ship ever built and only one of three such ships in the world (the other two being the German-built *Otto Hahn* and the Russian container ship *Sevmorput*).

1962

- *July 6:* The first experiment under Project Plowshare, code-named Sedan, is conducted in Nevada. Plowshare is a U.S. project for the use of nuclear weapons for large-scale excavation of earth.

1963

- *December 12:* The Jersey Central Power and Light Company announces it will build a nuclear power plant in Lacey Township, New Jersey. The plant is the first nuclear facility in the United States that is expected to provide an economical alternative to a fossil-fueled plant.

Chronology

1964

- *August 26:* President Lyndon B. Johnson signs the Private Ownership of Special Nuclear Materials Act, which allows the nuclear energy industry to own the fuel for its nuclear power plants. After June 30, 1973, private ownership of such fuels is mandatory.
- *October:* Three U.S. nuclear-powered surface ships—the NS *Enterprise*, NS *Long Beach*, and NS *Bainbridge*—complete a 30,565-mile (48,900-kilometer) round-the-world tour, during which they make no stops for refueling.
- *October 16:* China conducts its first test of a fission bomb at its Lop Nur Test Ground.

1965

- *April 3:* The first nuclear reactor is launched into space on board an Atlas Agena D rocket. The reactor operates successfully in space, generating more than 500 watts of power for 43 days.
- Scientists at the Brookhaven National Laboratory complete their revision of the WASH-740 report of 1957. Because of its anticipated negative effects about the safety of nuclear power among the public, the report is never published by the Atomic Energy Commission.

1966

- *October 5:* An accident at the Fermi I nuclear power plant outside Detroit causes the facility to shut down operations. Repairs are not completed until May 1970, but re-starting of the plant is delayed when a sodium explosion occurs within the reactor. The plant is closed permanently in August 1972.

1967

- *June 17:* China tests its first nuclear fusion bomb at its Lop Nur Test Ground.

1968

- *August 24:* France conducts its first test of a fusion bomb at Fangataufa Atoll in the South Pacific.

1969

- *October 29:* American researchers John Gofman and Arthur Tamplin present a paper at the Nuclear Science Symposium of the Institute of Electrical and Electronic Engineers (IEEE) arguing that the federal government's standards for maximum permissible radiation dose is about 10 times too high.

1970

- *December:* Construction begins at Hanford, Washington, on a Fast Flux Test Facility. The reactor is intended to be the prototype for the United States's breeder reactor program. This was expected to mark an important new step in the nation's nuclear power program, but the reactor operated for only 10 years before being decommissioned.

1971

- *July 23:* In *Calvert Cliffs' Coordinating Committee v. Atomic Energy Commission* (449 F.2d 1109), the U.S. Court of Appeals for the District of Columbia rules that decisions about the licensing of nuclear power plants by the Atomic Energy Commission are subject to the provisions of the National Environmental Policy Act of 1970. That act requires that an environmental impact statement be obtained for every new nuclear reactor facility, a provision that the AEC had argued did not apply to its own activities.

1973

- Power companies place orders for 41 new nuclear power plants, the largest number of such facilities ever ordered in a single year.

1974

- *May 18:* India conducts its first test of a fission weapon in the Rajasthan desert near the city of Pokaran.
- *August:* The Atomic Energy Commission releases a 14-volume report, "An Assessment of Accident Risks in U.S. Commercial Nuclear Power Plants." The report summarizes the results of a study conducted by an AEC committee chaired by Norman C. Rasmussen. The report is popularly known as "The Rasmussen Report," or the WASH-1400 report on reactor safety. The report suggests that there is very little risk to the general public as the result of an accident at a nuclear power facility. The AEC later repudiates many of the findings in the Rasmussen Report.
- *October 11:* President Gerald Ford signs the Energy Reorganization Act of 1974. The act divides the responsibilities and activities of the Atomic Energy Commission between a new Nuclear Regulatory Commission and an Energy Research and Development Administration.

- *November 13:* Karen Silkwood, an employee at the Kerr-McGee Cimarron plant near Crescent, Oklahoma, is killed in an automobile accident while apparently trying to deliver evidence of malfeasance at the plant to a representative of the Nuclear Regulatory Commission and a reporter for the *New York Times.*

1975

- *March 22:* A fire breaks out at the Brown's Ferry Nuclear Power Plant, a unit providing energy to the Tennessee Valley Authority. The fire begins when an electrical inspector at the plant uses a candle to check for air leaks in the reactor wall and sets fire to foam used to seal the leaks. The fire damages electrical cables in the reactor unit, causing the level of cooling water to drop to a dangerously low level. No lives are lost and no injuries to workers are reported, however.
- President Gerald Ford announces that the United States will forego the reprocessing of spent nuclear fuel produced from power plants. Prior to this time, it had been widely assumed that the primary method for dealing with nuclear wastes was to reprocess them, that is, convert them into materials that could be reused as fuels in nuclear power plants.

1976

- *July 11:* A group of individuals opposed to the construction of a nuclear power plant at Seabrook, New Hampshire, meets to form the Clamshell Alliance.

1977

- *April 30:* A group consisting of an estimated 18,000 nonviolent demonstrators, marching under the banner of the Clamshell Alliance, occupies the site of a proposed nuclear reactor at Seabrook, New Hampshire. About 1,400 protestors are arrested and jailed during the event.
- *August 4:* President Jimmy Carter signs the Energy Reorganization Act, creating the new Department of Energy (DOE) and transferring the responsibilities and activities of the Energy Research and Development Administration to the new department.
- *August 6:* An estimated 1,500 individuals led by the Abalone Alliance protest against the construction of a nuclear power plant at Diablo Canyon, California. The event is the first of many led by the alliance over the next nine years, although the plant is eventually built and licensed by the federal government.

- President Carter also reaffirms former President Gerald Ford's decision to renounce reprocessing as a method for treating waste materials produced by nuclear power plants. Instead, he announces an "away-from-reactor" program whereby nuclear wastes are to be transported to some distant site for disposal and storage.

1978

- *April 3:* In *Vermont Yankee Nuclear Power Corp. v. NRDC*, the U.S. Supreme Court rules that the environmental impact statement prepared in conjunction with the construction of a nuclear power plant is closely determined by the requirements set out by the Nuclear Regulatory Commission and are not superseded by the more general requirements of the Environmental Protection Act of 1970.
- *June 26:* In *Duke Power Co. v. Carolina Environmental Study Group*, the U.S. Supreme Court decides that, "in light of the extremely remote possibility of an accident" at a nuclear power plant, the Price-Anderson Act of 1957 is a reasonable method for providing compensation in case of such an accident and the act is not unconstitutional.
- *November 8:* The Uranium Mill Tailings Radiation Control Act takes effect. The act directs the Environmental Protection Agency (EPA) to develop methods for reducing the risk posed by radiation from the waste products of uranium mining.

1979

- *March 28:* The worst nuclear power plant accident in U.S. history occurs at the Three Mile Island nuclear power plant in eastern Pennsylvania when the emergency cooling system fails and approximately half the reactor core melts down, venting radioactive gases to the environment surrounding the plant. Contamination is so severe that clean-up activities do not begin until more than three years later. No deaths are attributed to the accident, although some experts believe that a number of workers and nearby residents may have suffered long-term health problems, such as increased rates of cancer.
- The Defense Authorization Act of 1979 (now Public Law 96-164) authorizes the Department of Energy to develop a research facility for demonstrating the safe disposal of radioactive waste produced by national defense activities. The site selected for this facility is located in the Chihuahuan Desert, 26 miles southeast of Carlsbad, New Mexico. The site, later named the Waste Isolation Pilot Project (WIPP), receives its first shipment of nuclear wastes on March 26, 1999.

Chronology

1980

- The U.S. Congress passes the Low-Level Radioactive Waste Policy Act, requiring every state to become responsible for all the low-level nuclear wastes generated within its borders.

1981

- President Ronald Reagan issues a "Nuclear Power Policy Statement" in which he asks for a larger role for nuclear power in the nation's energy equation and instructs the secretary of energy to develop as quickly as possible a reliable method for storing and disposing of commercial, high-level radioactive wastes.

1982

- The nation's oldest commercial nuclear power plant, at Shippingport, Pennsylvania, is shut down. Decommissioning of the plant is completed in 1989.
- The U.S. Congress passes the Nuclear Waste Policy Act (NWPA), establishing a policy and procedure for the disposal of the nation's commercial nuclear wastes. The act comes nearly 40 years after the "atomic age" has begun in the United States and is fated to encounter a long series of delays. President Reagan signs the act on January 7, 1983.

1983

- *April 19:* In the case of *Metropolitan Edison v. People v. Nuclear Energy*, the U.S. Supreme Court decides that the owner or operator of a nuclear power plant does not need to take into consideration emotional, mental, psychological, or similar nonphysical damage caused by the operation of the plant in its environmental impact statement.
- *October 26:* The U.S. Senate declines to provide further funding for the Clinch River breeder reactor project, effectively closing it down.
- Nuclear energy becomes the third most important source of electrical energy in the United States, surpassing natural gas for the first time in history. A year later, it passes hydroelectric power also, making it the second most important source of energy after coal.

1984

- *January 11:* In the case of *Silkwood v. Kerr-McGee Corp.*, the U.S. Supreme Court rules that a person can sue for damages resulting from an accident at a nuclear power plant under both state and federal law.

1985

- Congress passes the Low-Level Radioactive Waste Policy Amendments Act of 1985. The act strengthens and extends the provisions of the original act, passed in 1980, by encouraging states to form regional associations, called compacts, for the purpose of disposing of low-level radioactive wastes generated in their areas.

1986

- *April 25:* An accident occurs during routine safety tests on the reactor core of Unit 4 of the Chernobyl Nuclear Power Plant, near Kiev, Ukraine. The accident is by far the worst civilian nuclear disaster in history, with at least 4,000 deaths caused by radiation and as many as 10 million people across Europe exposed to dangerous levels of radiation.
- *December 9:* A pipe carrying superheated water breaks at the Surry Nuclear Power Plant in Virginia, releasing 30,000 gallons of very hot water. Eight workers are injured, four of whom later die. The pipes, originally a half-inch thick, had eroded in some places to about one-tenth of that thickness. The Surry accident was only one of two in the United States in which workers in a nuclear facility were killed, the other having been at the National Reactor Testing Site in 1961.
- The Perry Nuclear Power Plant, near Cleveland, Ohio, becomes the 100th nuclear power facility constructed in the United States.

1987

- *December 22:* The U.S. Congress approves legislation designating Yucca Mountain, Nevada, as the only site to be considered as a high-level nuclear waste repository.

1988

- The Price-Anderson Act of 1957 is reauthorized (P.L. 100-408). The 1988 amendments add two new provisions to the original act. First, indemnity coverage becomes required (it was optional under the original act). Second, provisions for monitoring the work done at nuclear power sites are strengthened to increase the safety of workers.

1989

- The decade of the 1980s sees the largest expansion of nuclear power in the United States to date, with 46 new facilities having been opened and the share of electrical power generated at such plants reaching just over 19 percent.

1990

- Consumers Power, a Michigan electric utility, decides to convert a planned nuclear power plant into a natural-gas cogeneration power plant. The company had already invested $4.2 billion in planning and construction on the nuclear power plant before deciding that such a plant would not be economical to complete and operate. The cogeneration plant now produces enough electricity for a city with a population of 1 million people and enough steam to power the Dow Chemical factory in Midland, Michigan.

1992

- The Energy Policy Act revises and streamlines the procedures by which nuclear power plants are licensed. The new provisions are described in the Code of Federal Regulations at 10 CFR Part 52.

1993

- *December:* The Department of Energy (DOE) announces that the Fast Flux Test Facility at Hanford, Washington, is to be shut down. A number of proposals are made for maintaining the facility for other purposes, such as using it as a commercial reactor, but none is accepted. DOE announces on December 19, 2001, that it will, in fact, shut down the reactor permanently.
- *December 9:* The Tokamak Fusion Test Reactor (TFTR) at Princeton University, for the first time in its operation, produces more energy than it consumes.

1994

- *June 17:* Attorneys for 14 utility companies and 27 state agencies from 20 states file suit with the U.S. Circuit Court of Appeals asking the court to affirm that the Nuclear Waste Policy Act require the Department of Energy to begin accepting spent nuclear fuel no later than December 31, 1993.
- *November 2:* The Princeton TFTR generates 10.7 million watts for a few seconds. That accomplishment sets a record so far unmatched for power production from a fusion reactor.
- The Nuclear Energy Institute (NEI) is formed from the merger of a number of older industry groups, including the American Nuclear Energy

Council, the U.S. Council for Energy Awareness, and the Nuclear Management and Resources Council. NEI's mission is to work for the adoption of policies that promote the beneficial uses of nuclear energy and technologies in the United States and around the world.

■ The Nuclear Regulatory Commission approves final design specifications for two new types of nuclear reactors, General Electric's Advanced Boiling Water Reactor (ABWR) and ABB Combustion Engineering's System 80+ Advanced Pressurized Water Reactor.

1997

■ *July 2:* The Department of Energy carries out the first of a series of underground subcritical tests on nuclear materials in the Operation Rebound program. A major purpose of the program is to determine the status of existing nuclear weapons and materials. The term *subcritical* refers to the fact that the tests are carried out in such a way that no nuclear chain reaction is possible. The tests continue until September 18, 1997.

■ President Bill Clinton asks his Committee of Advisors on Science and Technology to study energy needs in the United States in the 21st century. In its report, *Federal Energy Research and Development for the Challenges of the 21st Century*, the committee recommends an increase in the role played by nuclear energy in the nation's energy equation. Arising out of that recommendation, the president creates the Nuclear Energy Research Initiative, a program through which the federal government can support research by independent investigators on improvements in nuclear power production technology.

■ The number of unplanned automatic reactor shutdowns reported for the year drops to zero. This number is generally regarded as a measure of the safety of nuclear power plant operations by the industry.

1998

■ *April:* Baltimore Gas & Electric (BG&E) Company submits the first application for relicensing of a nuclear power plant. BG&E requests a 20-year extension of the license on Units 1 & 2 at Calvert Cliffs, Maryland. The Nuclear Regulatory Commission approves this application in March 2000. Later the same year, Duke Energy applies for license renewals for the three units of its Oconee plant, an application that NRC approves in May 2000.

■ *May 11:* India conducts its first test of a fusion bomb at its Pokaran test site.

■ *May 28:* Pakistan conducts its first test of a fission bomb in the Chagai region of Baluchistan province, near its borders with Iran and Afghanistan.

- *September 23:* The United States agrees to provide financial support for the International Thermonuclear Experimental Reactor (ITER) project for one year, although it has no plans to continue working with the project beyond that date.

1999

- *March 15:* A study conducted by the United Kingdom's National Radiological Protection Board finds that workers at nuclear facilities experience the "healthy worker effect," in which subjects tend to have fewer health problems than nonworkers in surrounding areas, presumably because of greater safety measures taken at the workplace. The results of the study are later criticized by a number of authorities in the field, primarily because of supposed procedural errors.
- *March 26:* The Waste Isolation Pilot Project in New Mexico receives its first shipment of transuranic wastes.
- *August 12:* French minister for the environment Dominique Voynet signs a decree authorizing the construction of an experimental underground storage site for high-level radioactive wastes at Bure, near the border of the departments of Heuse and Haute-Marne. The test site was authorized by the Nuclear Waste Law of 1991 that required the government to have in place by 2006 a system for disposing of the nation's spent fuel and other high-level radioactive wastes.
- *September 30:* An accident occurs at the Tokai-Mura nuclear power plant in Japan during which three workers are exposed to high levels of radiation. Two of the workers later die as a result of the accident. The Japanese government classifies the accident as a Level 4 incident, in which workers at the plant are exposed to serious risk from radiation, but no one outside the plant is considered to be at risk. The Tokai-Mura accident is later determined to be one of the five worst accidents to occur at a nuclear power plant to date.
- *October:* Burial of the complete reactor vessel from Portland (Oregon) General Electric's decommissioned Trojan Nuclear Power Station begins at a 100-acre site on the U.S. Department of Energy's Hanford Reservation. The vessel is 13 meters (43 feet) long and weighs about 1,000 metric tons (1,000 tons). The project is carried out by a nuclear waste disposal company, US Ecology, the first effort of its kind in history.

2000

- A long-term study of 32,135 individuals thought to have been at risk as a result of the 1979 accident at the Three Mile Island nuclear power plant by researchers at the University of Pittsburgh Graduate School of Public

Health finds no apparent increase in the rate of cancers in the experimental group compared to controlled groups not exposed to radiation during the same period of time.

- Three power companies—Entergy Operations, Southern Nuclear Operating Company, and Florida Power & Light—apply for license renewals of their nuclear power plants—Arkansas Nuclear One, Edwin I, Hatch 1 & 2, and Turkey Point 3 & 4, respectively.
- The average number of "significant events" per nuclear power plant in the United States drops to 0.03. A significant event is defined as an event that "challenges a plant's safety system." This number remains at or near this level ever since.

2001

- *March 30:* The Nuclear Regulatory Commission announces the formation of a Future Licensing Project designed to prepare for expected applications from industry for the construction of new nuclear power plants within the next few years.
- *May 18:* President George W. Bush's National Energy Policy strongly recommends increased attention to nuclear power as a source of energy in the nation's future plans. It makes a number of specific recommendations designed to achieve this objective, including encouraging the Nuclear Regulatory Commission to facilitate industry efforts to expand the number of nuclear power plants in operation in the United States, urging the commission to extend licenses on existing plants, and suggesting the need for legislation that would transfer some of the financial risk of nuclear power plant construction and operation from industry to the federal government.
- *July 6:* Loyola De Palacio, Energy Commissioner and Vice-President of the European Commission, presents a strong argument for maintaining, if not increasing, the role of nuclear power in the generation of electricity in Europe. " It would be imprudent to renounce nuclear energy," she says. "Without nuclear [power] Europe would not be capable to comply with the Kyoto Protocol requirements [for reducing greenhouse gas emissions]."
- *July 27:* H.R. 4, commonly known as the Securing America's Future Energy Act of 2001, is introduced into the U.S. Congress, with a number of provisions included to encourage the growth and development of nuclear power in the United States. Among these provisions are approximately $2.5 billion in tax breaks and subsidies to the nuclear industry.
- *December 14:* The Federal Bureau of Investigations (FBI) reports on an investigation of the safety of the Indian Point Nuclear Power Plant, over

which one terrorist-hijacked plane flew on September 11, 2001. The report finds that the plant is "an extremely safe place," and that residents in the surrounding area should have no fears of serious damage to the plant in case of a terrorist attack.

2002

- *March 6:* Workers at the Davis-Besse Nuclear Power Station in Ohio discover a hole in the reactor vessel head about the size of a football. The hole has apparently been caused by the reaction between boric acid, produced within the reactor, and the metal of which the reactor head is made. Critics claim that the hole could have resulted in the loss of cooling water, resulting in a Three Mile Island–type accident at the plant.
- *March 8:* The U.S. Senate approves a bill to extend the Price-Anderson Act.
- *June 15:* An earthquake of magnitude 4.4 on the Richter scale strikes in the Yucca Mountain region planned for the nation's nuclear waste depository site. Officials point out that the site is designed to withstand an earthquake with 30,000 times more energy and argue that the event poses no threat to plans for construction of the site.
- *July 24:* President George W. Bush signs a resolution approving Yucca Mountain, Nevada, as the site of the nation's repository for spent fuel and high-level radioactive wastes. The resolution follows Senate approval of the Yucca Mountain site a month earlier, allowing the Department of Energy to submit a license application with the Nuclear Regulatory Commission for construction to begin on the site.
- *September 10–12:* A conference on The Nuclear Renaissance is held in Washington, D.C., sponsored by Atomic Energy of Canada Limited (AECL), Excel Services Corporation, Framatome ANP, and Winston & Strawn. At the conference, industry representatives and representatives of the U.S. government optimistically survey the outlook for nuclear power in the coming generation.
- *September 12:* A conference committee of the U.S. Senate and House agrees to reauthorize the Price-Anderson Act for one year, to August 1, 2003. The act officially expired on August 1, 2002. Congress then begins debate on an extension of the act for 15 more years, to 2017.

2003

- *February 1:* President George W. Bush announces that the United States will rejoin the International Thermonuclear Experimental Reactor (ITER) project, a program that it left in 1999.

- *February 13:* The U.S. Department of Energy creates a new Office of Legacy Management to take responsibility for long-term care of sites formerly used for the production of nuclear weapons.
- *September 4:* The World Nuclear University is opened in London. Supported by a number of national and international nuclear groups, the university's mission is to promote research and education that will further the development of peaceful applications of nuclear energy throughout the world.
- *November 14:* Germany shuts down the 32-year-old Stade nuclear power plant near Hamburg. The closure is the first step in the government's announced plans to decommission all 19 of its nuclear power plants by the year 2025. The government has announced no plans for replacing the energy obtained from the plants.
- During the year, the Nuclear Regulatory Commission renews the licenses of five nuclear power plants, bringing to 10 the number of plants that have been relicensed under the more liberal licensing policies established by the administration of President George W. Bush. Provisions under which relicensing occurs are summarized in the Code of Federal Regulations, 10 CFR Parts 51 and 54.

2004

- *June 6:* A poll conducted by the Nuclear Energy Institute finds that 65 percent of Americans interviewed agree that nuclear power should be part of the nation's overall energy equation for the future. Seventy-two percent of those interviewed thought that the nearest nuclear power plant was safe and reliable.
- *July 9:* The Federal Appeals Court for the District of Columbia rules that the Department of Energy's plans for storing nuclear wastes at Yucca Mountain, Nevada, for a period of 10,000 years are inadequate. The court directs the department to amend its plan to provide for safe storage of such materials for an even longer period of time, although the court does not specify how long that period is to be.
- *August 8:* The U.S. Department of Energy signs an agreement with the French atomic energy commission to allow cooperation between the two nations that will allow the United States to take advantage of fast-breeder technology developed in the French Phenix project, a technology that is no longer available in the United States.
- *September 14:* The Nuclear Regulatory Commission approves a new nuclear reactor design, the Westinghouse AP 1000, for which interest has already been expressed by companies in Asia, Europe, and the United States.

2005

- *January 1:* The World Nuclear Association announces that 25 new nuclear power plants are under construction around the world, that 37 additional plants have been ordered, and that 74 more plants have been proposed. The greatest growth takes place in India, where nine new plants are being built and 24 more planned, and in China, with four plants under construction and 24 in the planning stages.

- *January 30:* Great Britain's Labour government issues a white paper announcing it intends to push for a significant increase in nuclear power plant production should it be reelected in forthcoming elections.
- *April 18:* The Indonesian government gives approval for the construction of the nation's first nuclear reactor on the island of Java. The reactor is expected to begin producing electricity in 2016.
- *April 24:* Officials from more than 50 nations meet in Vienna for a conference sponsored by the International Atomic Energy Agency to discuss ways to prevent catastrophic accidents such as the one that occurred at the Chernobyl power plant in 1986.
- *April 26:* President George W. Bush outlines his energy plans for his second presidential term, stating that nuclear power must play an essential role in the nation's energy future and asking Congress to pass legislation protecting the nuclear industry in case of a catastrophic accident.
- *May 11:* The government of North Korea announces that it has removed 8,000 fuel rods from a nuclear power plant at Yongbyon for use in the construction of weapons to add to its existing nuclear arsenal. With no outside observers present in the North Korea, the status of that nation's nuclear weapons program remains a mystery.
- *May 18:* In spite of heavy pressure from the European Union and the United States, the government of Iran announces that it intends to go ahead with plans to resume work on its national nuclear power program. Opposition to Iranian plans are based on fears that materials used in the nuclear program could be diverted for construction of nuclear weapons.
- *July 12–15:* Delegates from Australia, Canada, China, France, Hong Kong, South Korea, Malaysia, Singapore, Thailand, the United States, and Vietnam gather in Hong Kong for the Nuclear Power Asia Pacific 2005 conference, aimed at exploring the role of Asian nations in promoting the use of nuclear power for the purpose of transforming the global energy landscape.
- *July 20:* President George W. Bush announces that the United States will assist the Indian government in the development of its nuclear reactor development program. The action is the first instance in which the U.S. government has offered such support to a nation that has not agreed to outside monitoring of its nuclear facilites. The president's decision arouses concern among officials in many nations, including India's traditional rival, Pakistan.
- *August 8:* President George W. Bush signs the 2005 energy bill, which, among other provisions, allots $1.5 billion in direct subsidies to private industries for the construction of new nuclear power plants.

CHAPTER 4

BIOGRAPHICAL LISTING

This chapter contains brief biographical sketches of individuals who have played major roles in the development of nuclear energy in the United States and throughout the world.

Clinton Presba Anderson, Democratic senator from the state of New Mexico from 1948 to 1972, following three terms in the U.S. House of Representatives. He was coauthor, along with Congressman Melvin Price (D-Ill.), of the Price-Anderson bill in 1957 that limited the liability of private companies in case of an accident at a nuclear power plant.

Francis William Aston, a British physicist who was awarded the 1922 Nobel Prize in chemistry for his discovery of isotopes. Isotopes are different forms of an element with the same atomic number but different atomic masses. Although they have identical chemical properties, they differ in other essential ways, such as their nuclear properties. As an example, two isotopes of uranium, uranium-233 and uranium-235, undergo fission, while a third (and much more abundant) isotope, uranium-238, does not.

Bernard Mannes Baruch, a partner in the investment firm of A. Houseman who made his fortune before the age of 30. Baruch served as an economic adviser to Democratic presidents Woodrow Wilson, Franklin D. Roosevelt, Harry S. Truman, and John F. Kennedy. He served during World War I as chair of the War Industries Board, represented Wilson at the Versailles Peace Conference, and was a member of Roosevelt's "Brain Trust." In 1946, Truman appointed Baruch to represent the United States in the United Nations Atomic Energy Commission, where he presented a plan for the international peacetime control of atomic energy. The plan was later rejected.

Antoine-Henri Becquerel, a French physicist whose discovery of radioactivity in 1896 earned him the 1903 Nobel Prize in physics. Becquerel's research was among the earliest investigations that showed the relationship between matter and energy.

Hans Albrecht Bethe, German theoretical physicist. Born in 1906, he left his homeland in 1933 with the rise of the Nazi Party. Bethe was especially interested in mechanisms by which the stars produce energy. In 1939, he proposed the so-called carbon-nitrogen cycle as a possible explanation for this phenomenon, an accomplishment for which he was awarded the 1967 Nobel Prize in physics. Bethe's theories were of critical importance in the design and production of the first fusion bombs.

Aage Niels Bohr, a Danish physicist who proposed the nuclear model of the atom in 1913. According to this model, atoms consist of a central core that is positively charged, surrounded by shells of negatively charged electrons. Although it has since been greatly refined, the Bohr model continues to serve as a general picture of the composition of atoms. Bohr was awarded the 1922 Nobel Prize in physics for his study of atomic structure.

Vannevar Bush, an electrical engineer with a doctorate from the Massachusetts Institute of Technology. He became interested and involved in the administrative aspect of scientific research and in government service in the late 1930s. Bush was chosen by President Franklin D. Roosevelt to serve as head of the National Defense Research Committee during World War II and, in October 1941, he met with President Roosevelt and Vice President Henry Wallace to argue for the development of a fission bomb by the United States. Largely as a result of his efforts, the president created the Manhattan Project, in which Bush continued to have a modest role for a number of years.

Helen Caldicott, an American physician. She has campaigned against both fission weapons and nuclear power plants for more than a half century. She is the founder of an antinuclear group known as Physicians for Social Responsibility and has written several books, including *Nuclear Madness, Missile Envy, A Desperate Passion,* and *The New Nuclear Danger.* She has said that "nuclear power, apart from nuclear war, is the greatest medical threat posed to life on this planet."

James Chadwick, an English physicist. He carried out experiments in the early 1930s that eventually led to the discovery of the neutron, an uncharged particle of about the same mass as that of the proton located in the nucleus of all atoms (except hydrogen). In 1948 he was appointed to the British mission to the Manhattan Project, where he developed a close working relationship with General Leslie Groves. After the war, Chadwick served as an adviser to the research team that built the first British fission bomb.

Joan Claybrook, consumer advocate who cofounded with Ralph Nader the public interest group Public Citizen. She also co-organized the first national conference in opposition to nuclear power plants, Critical Mass

'74, and a group by the same name within Public Citizen. She later served as president of Critical Mass and of the consumer advocacy agency Congress Watch. From 1977 to 1981, she was head of the National Highway Traffic Safety Administration.

Bernard L. Cohen, professor of chemistry, radiation health, and environmental and occupational health at the University of Pittsburgh for 36 years. He has been an outspoken advocate for the development of nuclear power in the United States and around the world. Cohen is best known for his argument that the widely accepted linear–no threshold dose–response theory of radiation effects is untrue. According to that theory, there is no level of radiation that does not produce at least some deleterious effect on humans and other animals. Cohen conducted a number of studies suggesting that the health effects of low-level radiation are much less severe than those accepted by many specialists in the field.

William Sterling Cole, Republican congressman from New York State. He was coauthor (along with Senator Bourke Hickenlooper) of the Atomic Energy Act of 1954. Cole served in the House of Representatives from 1935 to 1957 and was chairman of the Joint Committee on Atomic Energy from 1953 to 1954. In 1957, he was appointed Director General of the International Atomic Energy Agency, headquartered in Vienna, Austria, a post he held until 1961.

Arthur Holly Compton, an American X-ray physicist who won the Nobel Prize in Physics in 1927 for his discovery of the so-called Compton Effect. He was appointed chair of the National Academy of Sciences Committee to Evaluate Use of Atomic Energy in War in 1941, a committee whose recommendations led to the formation of the Manhattan Project. From 1942 through 1945, he was director of the Chicago Metallurgical Laboratory, where much of the fundamental research on nuclear reactors was conducted.

James Bryant Conant, an American chemist and long-term president of Harvard University. He served as head of the National Defense Research Committee, which oversaw fundamental research on which the construction of the first fission weapons was based. After the war, he served as U.S. High Commissioner to Germany and ambassador to Germany.

Marie Curie, a Polish-French physicist who won two Nobel Prizes. The first prize, in physics, was awarded in 1903 for her discovery (along with her husband Pierre) of natural radioactivity. The second prize, in chemistry, was awarded in 1911 for her discovery of the elements radium and polonium. Curie served as Professor of General Physics at the University of Paris (Sorbonne) from 1904 through 1934.

Pierre Curie, a French physicist who was awarded a share of the 1903 Nobel Prize in physics, along with his wife, Marie, for their discovery of radioactivity. He was Titular Professor of Physics at the Sorbonne from 1904 until his death two years later in a horse-cab accident.

John Dalton, an English chemist and physicist. He is generally regarded as the father of the modern atomic theory. Dalton based his theory on careful measurements of the masses of elements that combine with each other in the formation of compounds. In addition to his work in atomic theory, Dalton made important contributions to meteorological studies, the study of color-blindness, the behavior of gases, and a variety of other subjects.

Democritus of Abdera, a Greek natural philosopher. He is thought to have lived between about 370 and 460 B.C. Although little is known of his life, his contributions to philosophical thought are relatively well known. He taught that matter consists of tiny particles, which he named *atomos* ("indivisible"), that are eternal, invisible, so small that their size cannot be determined or reduced, incompressible, and differing from each other only in size, shape, arrangement, and position.

Carrie Dickinson, a farm wife, former school teacher, registered nurse, and owner/operator of Aunt Carrie's Nursing Home in Claremore, Oklahoma. She became an antinuclear activist at the age of 56 in 1973 when she read about plans for the construction of a nuclear power plant in nearby Inola. She worked for nine years, expending nearly all her life's savings, to prevent construction of the plant, a battle she eventually won. She later became involved in efforts to prevent the dumping of nuclear wastes on lands owned by Native American tribes.

Albert Einstein, a theoretical physicist born in Germany who emigrated to the United States in 1933. He is generally regarded as one of the half dozen greatest scientists who ever lived. His most productive year was 1905, when he published three famous papers on Brownian movement, the quantum nature of light, and the special theory of relativity. It was the last of these papers that elucidated the now-famous relationship between mass and energy ($E = mc^2$) on which all nuclear weapons and peacetime applications are based. In 1939, Einstein was persuaded by a number of his colleagues to sign a letter to President Franklin D. Roosevelt outlining the potential use of nuclear energy for the development of weapons.

Dwight David Eisenhower, 34th president of the United States. He took office on January 20, 1953, at a crucial point in the early history of nuclear energy in the United States, only 13 days after outgoing president Harry S. Truman had announced that the United States had been successful in developing a fusion bomb. Eisenhower was faced immediately, therefore, with the early stages of a "cold war" battle between the United States and the Soviet Union in which each nation was committed to building as large and as powerful a nuclear arsenal as possible. At the same time, Eisenhower was confronted with a host of promises about the peacetime applications of nuclear energy that, supporters claimed, would transform the modern world. Arguably the most public manifestation of Eisenhower's interest in nuclear issues was his Atoms for Peace plan, announced on December 8, 1953, in front of the United Nations General Assembly. Under this plan, Eisenhower offered to make available to countries around the world nuclear information possessed by the United States for the purpose of developing peacetime applications of the technology.

Enrico Fermi, an Italian-American physicist. He made a number of important discoveries that contributed to the early development of nuclear science. Fermi was that somewhat rare individual who was equally at home in dealing with theoretical and experimental research. One of his earliest discoveries, for example, was a set of statistical laws (now known as Fermi statistics) that determine the behavior of fermions, a class of particles also named after him. By contrast, the Nobel Prize in physics awarded to him in 1938 was to honor his discoveries of the way new isotopes are produced as a result of n,γ reactions and the mechanism by which nuclear fission is brought about by slow neutrons. During World War II, Fermi directed the experiments carried out at the Chicago Metallurgical Laboratory that resulted in the development of the first nuclear reactor.

Daniel Ford, an American economist. He joined with Henry Kendall, then professor of physics at the Massachusetts Institute of Technology, to conduct a study in 1971 of the safety of the nuclear reactor at the Pilgrim Nuclear Power Station in Plymouth, Massachusetts. The two concluded that the plant's emergency system was inadequate to protect against certain kinds of accidents that could result in meltdown of the reactor core. Ford later became executive director of the Union of Concerned Scientists and wrote and coauthored a number of books on nuclear issues, including *An Assessment of the Emergency Core Cooling Systems Rulemaking Hearing* (1974), *Beyond the Freeze: The Road to Nuclear Sanity* (1982), *Browns Ferry: The Regulatory Failure* (1976), *The Cult of the Atom: The Secret Papers of the Atomic Energy Commission* (1982), *Meltdown* (1986), and *The Nuclear Fuel Cycle: A Survey of Public Health, Environmental, and National Security Effects of Nuclear Power* (1974).

Otto Robert Frisch, an Austrian-English physicist. He is best known for his collaboration with his aunt Lise Meitner in the interpretation of the first atomic fission reactions in 1938. As was the case with many German and Italian scientists during the 1930s, Frisch left Austria in 1933 to avoid Nazi persecution of Jews. He was living in London when he received word of the historic experiments carried out by Otto Hahn and Fritz Strassman in which uranium atoms were split by neutrons. He then worked with Meitner to develop a mathematical explanation of the observations made by Hahn and Strassman.

Richard L. Garwin, American physicist. He has been interested in policy issues related to nuclear science for many years. Garwin has long been in the somewhat interesting position of working on the development of nuclear weapons and also on efforts to bring about control over such weapons. In a 2001 interview with the *New York Times,* for example, Edward Teller, the so-called Father of the Hydrogen Bomb, credited Garwin with having carried out the calculations that made such a weapon possible. Garwin himself agreed, saying that, by the time the Manhattan Project had been completed, everyone involved in the project was "burnt out." Since the calculations Teller wanted were "the kind of thing I do well," Garwin said, he just went ahead and did the needed calculations, making him, if not the Father of the Hydrogen Bomb, at least its midwife. Yet, Garwin has always been conflicted about his contributions to the development of nuclear weapons. "If I could wave a wand," he has said, to make the fusion bomb and the whole nuclear age just disappear, "I would do that." To that end, Garwin has served on a number of committees working to control the use of nuclear energy, for which he was awarded (among other honors) the 1991 Ettore Majorana Erice Science for Peace Prize.

John Gofman, a physician and nuclear chemist. He is famous for his efforts, with colleague Arthur Tamplin, to have a reduction made in the maximum permissible dose of radiation received by an individual. As a result of their studies in the 1960s, Gofman and Tamplin became convinced that standards established by the Atomic Energy Commission for radiation dosage was as much as 10 times too high. Although they battled for a number of years for this position, they were unable to convince the AEC and, in the process, lost their jobs and federal grants and suffered other forms of retaliation from the federal government.

Leslie R. Groves, a colonel in the U.S. Army in the early 1940s and later military commander of the Manhattan Project, the research program whose objective it was to develop the first nuclear weapons. Groves was promoted to the rank of brigadier general with his appointment to the Manhattan Project and was selected because he had considerable experience with construction projects and had a reputation as an above-average administrator. Although he was not a particularly well-liked individual, he was able to direct the project to a successful conclusion. After the war ended, Groves conducted a vigorous campaign to keep control of atomic energy within the military, a campaign that he eventually lost with the creation of the civilian-based Atomic Energy Commission.

Otto Hahn, a German physical chemist. He carried out one of the most important and essential experiments in the history of nuclear science. He and his coworker, Fritz Strassman, demonstrated in the late 1930s that the collision of a neutron with the nucleus of a uranium atom resulted in the formation of two new atomic nuclei with roughly equal masses; that is, that the neutron had split the uranium atom. Hahn and Strassman were so surprised by their results that they essentially refused to accept their implication. The correct interpretation of the experiment was provided, instead, by Austrian-Swedish physicist Lise Meitner and Austrian-English physicist Otto Frisch.

Bourke Blakemore Hickenlooper, Republican senator from Iowa who served in the U.S. Senate from 1944 to 1968. He was one of the coauthors (along with Congressman Sterling Cole [R-N.Y.]) of the Atomic Energy Act of 1954. Hickenlooper became chair of the Joint Committee on Atomic Energy (succeeding Senator Brien McMahon) when Republicans won control of Congress in 1946. Throughout his term of service, Hickenlooper led the business-oriented Republican campaign for the promotion of private use of nuclear energy, often placing him at odds with Democratic colleagues on the Joint Committee and in Congress.

Chester Earl Holifield, Democratic congressman from California for 32 years, from 1943 to 1974. During his first term in office, he developed a profound interest in nuclear energy to the extent that he eventually became known as "Mr. Atomic Energy" to many of his colleagues. Holifield was a strong supporter of the development of a fusion bomb and for many years carried on a campaign to ensure that the benefits of nuclear energy were made available as widely as possible to the general public rather than to private industry alone. A bill he sponsored for this purpose in 1956 with Senator Albert Gore (father of future vice president Al Gore) of Tennessee failed to pass, and similar efforts met the same fate as long as the Republican Party controlled Congress.

Edwin Carl Johnson, Democratic senator from Colorado, who served from 1937 to 1955. In 1945, he coauthored (with Representative Andrew J. May [R-Ky.]) a bill to place responsibility for the postwar development of atomic energy in the hands of the U.S. military. Although widely expected to pass easily, the May-Johnson bill encountered unexpected opposition from scientists, who preferred a civilian agency to have control over nuclear issues. The bill was eventually abandoned in favor of the McMahon bill, which created a civilian Atomic Energy Commission to supervise the development of nuclear energy in the United States.

Henry Kendall, American physicist and winner of the Nobel Prize in physics for 1990. He was one of the founders of the Union of Concerned Scientists (UCS) in 1969. UCS was formed by an ad hoc group of 50 senior faculty at the Massachusetts Institute of Technology who had become concerned about the intrusion of the U.S. military in a number of civilian and academic fields. Kendall was a rare individual who was able to conduct exciting breakthrough research in particle physics while remaining immersed in a variety of socioscientific issues. He served as executive director of UCS, testified before the U.S. Congress on a variety of scientific and social issues, and wrote a number of books on nuclear energy. In 1992, he wrote a "World Scientists' Warning to Humanity," later signed by about 1,700 of the world's leading scientists, in which he warned that "Human beings and the natural world are on a collision course."

Martin Heinrich Klaproth, a German chemist and the discoverer of uranium in 1789. The son of a tailor, at the age of eight Klaproth and his family were thrown into poverty when another family destroyed their home. He was apprenticed to an apothecary and, over the years, became a self-taught chemist. He is probably best known today for his discovery of new elements. He was among the first to recognize the elemental character not only of uranium but also of the elements zirconium (1789), titanium (1792), strontium (1793), chromium (1797), and cerium (1803).

Willard F. Libby, an American chemist and the first chemist ever appointed to the Atomic Energy Commission. He served on the commission from 1954 to 1959 before resigning to become professor of chemistry at the University of California at Los Angeles (UCLA). He was later appointed Director of the Institute of Geophysics and Planetary Physics at UCLA. While at the AEC, Libby was head of President Dwight D. Eisenhower's Atoms for Peace project. Libby was awarded the 1960 Nobel Prize in chemistry for his discovery of the carbon-dating method.

David Lilienthal, an attorney who earned his law degree at Harvard University. He spent most of his life in public service. He served for several years on the Wisconsin Public Service Commission and was appointed a director of the Tennessee Valley Authority (TVA) in 1933. After many years with that organization, he was chosen by President Harry S. Truman in 1946 to become first chairman of the Atomic Energy Commission. During his confirmation hearings, Lilienthal found himself at odds with almost every major figure in Washington over the future of atomic energy. He believed that nuclear issues should be controlled by civilian agencies, while most politicians and military leaders were arguing for military control of the new technology. Although Lilienthal's position carried the day with the passage of the McMahon bill in 1946, he continued to battle with his opponents over the relative importance of military and peacetime applications of nuclear energy for the rest of his life.

Amory Lovins, American physicist. He was trained at Harvard and Oxford and is widely respected for having changed the fundamental parameters of the debate over energy usage in the 1970s. Lovins argued that the world's dependence on fossil fuels and nuclear power would eventually (and sooner than later) have to be replaced by a greater use of alternative sources of energy, such as solar, wind, geothermal, and biomass. At a time when nuclear power was becoming increasingly popular in the United States and the rest of the world for the generation of electrical energy, Lovins was calling for a halt on construction of new nuclear power plants and a dismantling of existing plants.

Andrew J. May, Republican congressman from Kentucky for 16 years. He was coauthor (with Senator E. C. Johnson) of the May-Johnson bill introduced in October 1945. The May-Johnson bill represented the U.S. military's position on the post–World War II future of atomic energy, namely that the military should be largely responsible for decisions as to how nuclear energy was to be utilized. Although expected to pass easily in the Congress, the bill met substantial opposition among scientists and was later rejected in favor of the McMahon bill, granting authority over nuclear energy issues to civilian agencies. May was convicted in 1947 of accepting bribes during World War II and served nine months in prison.

Brien McMahon, Democratic senator from Connecticut who served two terms, the second of which was cut short by his death in 1952. McMahon was at the core of the debate at the end of World War II as to how nuclear energy should be controlled in the United States. In his role as supporter of civilian control, he battled military leaders, legislators, and government officials who thought the military should be responsible for whatever developments in nuclear energy might take place. The debate over this issue was one of the longest and most contentious in the history of the U.S. Congress and was resolved only on August 1, 1946, when Congress passed the McMahon bill, later to be known as the Atomic Energy Act of 1946.

Lise Meitner, an Austrian-Swedish physicist. She was primarily responsible (along with her nephew Otto Frisch) for explaining the results of the first experiment in which a uranium atom was split by her colleagues Otto Hahn and Fritz Strassman. Although Hahn and Strassman carried out the original experiments in which fission occurred, they were unable to interpret their results successfully. Only after Meitner and Frisch analyzed the Hahn-Strassman data were the astounding implications of the research made clear. Although Hahn was later awarded the Nobel Prize in physics (1944) for this research, Meitner's contributions were never recognized by the Nobel commitee. She was, however, awarded a share of the first Fermi Award, given by the U.S. Atomic Energy Commission in 1966.

Gregory C. Minor, Richard B. Hubbard, and Dale G. Bridenbaugh, engineers at the General Electric Reactor Division who resigned from their jobs in 1976 in protest against what they saw as blatant disregard for public safety by the federal government and the nuclear industry itself. The three men later formed a consulting firm, MHB Technical Associates, that carried out studies on the safety, reliability, construction, and economics of power plants. They also served in 1979 as consultants to the filming of a motion picture, *The China Syndrome*, that presented a worst-case scenario as to what might happen in the event an accident occurred at a nuclear power plant.

Ralph Nader, a consumer advocate who organized the first national conference of opponents to nuclear power, Critical Mass '74, held in Washington, D.C., in 1974. One outcome of the conference was the founding in the same year of the group Critical Mass within Nader's advocacy organization, Public Citizen. Nader and his colleagues held a similar meeting of antinuclear activists in 1975.

George Norris, a relatively little known nuclear physicist and attorney. He is credited with authorship of the Atomic Energy Bill of 1954, sponsored in Congress by Senator Bourke Hickenlooper and Representative Sterling Cole. Norris is said to have been a fervent advocate of the development of peacetime applications of nuclear energy, but he was frustrated by early governmental restrictions (such as those contained in the Atomic Energy Act of 1946) on the release of nuclear information to industry. He apparently wrote the first draft of the 1954 bill and then worked with Hickenlooper, Cole, and other congressional supporters to shepherd the bill through Congress.

J. Robert Oppenheimer, an American physicist who was widely recognized as one of the most gifted physicists ever produced in the United States. During World War II, he served as director of the atomic bomb project research being conducted at Los Alamos, New Mexico. After he left Los Alamos, he continued to serve the government as chair of the General Advisory Committee of the Atomic Energy Commission from 1945 to 1952. As a result of his experience in bomb research, Oppenheimer became convinced that scientists had a responsibility to do everything possible to prevent nuclear weapons from ever being used again. As a result, he strongly opposed U.S. efforts to construct a fusion bomb. In taking that position, he made a number of enemies among scientists, politicians, and government officials who felt that a fusion bomb was essential to the future defense program of the United States. Some of these individuals eventually raised questions as to Oppenheimer's patriotism and were successful in having his security clearance withdrawn, preventing him from carrying out further research on nuclear energy. Oppenheimer was chosen as director of the Institute of Advanced Studies at the end of World War II in 1947, a post he held until 1966, a year before his death.

Michael Phelps, an American chemist. He invented the procedure now known as positron emission tomography (PET) in 1973. PET scanners are now among the most common noninvasive methods of diagnosis used in medicine. Phelps has been on the faculty at the Washington University School of Medicine (1970–75), the University of Pennsylvania (1975–76), and the University of California School of Medicine (1976 to the present date). In 1998, Phelps was awarded an Enrico Fermi Award for his invention of the PET procedure.

Robert Pollard, a nuclear reactor engineer who resigned from his job at the Nuclear Regulatory Commission in 1976 because he had become convinced, as he later said, that the NRC was "more interested in protecting the nuclear industry than the health and safety of the public." After he left the agency, he joined the Union of Concerned Scientists (UCS), an organization formed in 1969 to "combat the perceived threat of the United States military's spread onto campuses, throughout southeast Asia, and into space." For more than a dozen years, Pollard served as UCS's specialist on matters of nuclear energy. Pollard has written, coauthored, and edited three books on nuclear energy: *The Nugget File: Excerpts from the Government's Special File on Nuclear Power Plant Accidents* (1979), *O-Rings and Nuclear Plant Safety: A Technical Evaluation* (1986), and *Three Mile Island—Have We Reached the Brink of Nuclear Catastrophe?* (1979).

Charles Melvin Price, Democratic representative from Illinois from 1945 until his death in 1988. Price was coauthor, along with Senator Clinton Anderson (D-N.M.), of the Price-Anderson bill in 1957 that restricted the liability of private companies in case of an accident at a nuclear power plant. Price served as chair of the Joint Committee on Atomic Energy from 1973 to 1975.

Dorothy Purley, a member of the Native American Pueblo tribe who worked for many years as a truck driver at the world's largest open pit uranium mine, near her home at the Laguna Pueblo in New Mexico. When she learned in the late 1990s of the radiation hazard to which she, her family, and her neighbors had been exposed for more than 60 years, she became an outspoken opponent of nuclear power development, in general, and of uranium mining, in particular. She died of cancer, thought to have resulted from exposure to radiation at the uranium mine, on December 2, 1999.

Dixie Lee Ray, first woman chair of the Atomic Energy Commission. She served in that position from 1973 until the agency was abolished and replaced by the Nuclear Regulatory Commission. Ray was trained as a marine biologist and had served as governor of Washington State from 1977 to 1981, having been elected as an independent. Ray was an avid supporter of nuclear power and frequently ridiculed those who expressed concerns about the safety of nuclear power plants. After leaving the AEC, she pursued an ongoing campaign against what she saw as an irrational concern for nature by environmentalists. She wrote two books expressing her views on this topic, *Environmental Overkill* (1994) and *Trashing the Planet* (1990).

Hyman George Rickover, admiral in the U.S. Navy. He was the most vigorous advocate for the development of nuclear vessels in the late 1940s and early 1950s. Although his views were in conflict with nearly all his superiors and most members of the U.S. Congress, the Atomic Energy Commission, and the Joint Committee on Atomic Energy, Rickover was finally successful in seeing the accomplishment of his dreams. Largely as a result of his efforts, the first of a number of nuclear submarines, the N.S. *Nautilus*, was launched on January 21, 1954.

Bertram Roberts, a former employee of the Florida Power and Light Company (FPL). He developed myelogenous leukemia in 1993 after retiring from his job. He and his wife, Hanni, then filed suit against FPL claiming that Bertram's cancer had resulted from exposure to radiation on the job site. The Robertses' claim was later denied both by the U.S. District Court for the Southern District of Florida and, upon appeal, by the U.S. Eleventh Circuit Court of Appeals. Both courts pointed out that the Roberts had not claimed, nor had they demonstrated, that Bertram had ever been exposed to radiation at a level greater than the federal government's standard.

Wilhelm Conrad Roentgen, a German physicist. He is best known for his discovery of X-rays in 1895. While studying the emissions produced when electrical current is passed through highly evacuated glass tubes, Roentgen noticed a screen at the far end of his room beginning to glow. He was able to demonstrate that the glow was called by radiation—later called X-rays—released from the tubes. The discovery was important because it demonstrated the close relationship between matter and radiation. For his discovery, Roentgen was awarded the 1901 Nobel Prize in physics.

Franklin Delano Roosevelt, 32nd president of the United States. As president, he was placed in a position of having to make what was arguably the most critical military decision during World War II when he authorized research on the development of a fission bomb. A letter, written to Roosevelt by Albert Einstein and Leo Szilard on August 2, 1939, and which discussed the merits of nuclear research languished in the White House for some time before the president was convinced by close advisers Vannevar Bush and James Bryant Conant of the importance of going forward with nuclear research. On October 9, 1941, Roosevelt finally gave his approval for the initiation of this program, giving it the code name Manhattan Engineering District (it was later renamed Manhattan Project). Roosevelt died before a nuclear bomb was ever tested, and the decision as to whether, when, and how the weapon was to be used was left to his successor, Harry S. Truman.

Nuclear Power

Lord Ernest Rutherford, one of the greatest experimental physicists of all times. He was born in New Zealand but spent most of his adult life in England and Canada. In 1899, he discovered and named alpha and beta particles, which are released when atoms are bombarded with radiation. Only a year later, he found that radioactive materials decay according to a characteristic and identifiable pattern, which he named their half-life. In 1911, he carried out an experiment for which he is particularly famous, the gold foil experiment, the results of which allowed him to announce that atoms are composed of two major parts: a dense, massive core, which he named the nucleus, and a collection of light, negatively charged electrons that orbit around the nucleus. Rutherford was awarded the 1908 Nobel Prize in chemistry and, perhaps of equal significance, he was involved in the training of 11 other scientists who themselves became Nobel laureates.

Andrei Sakharov, a Soviet physicist. He worked on the Soviet hydrogen bomb project from 1948 until 1968, when he was removed from the project because of ideological differences with the government about the use of fusion power. Sakharov and his longtime mentor Igor Tamm invented a device known as a tokamak for the controlled release of energy produced by fusion reactions.

Karen Silkwood, an employee at a plutonium production facility in Crescent, Oklahoma. She was killed in a suspicious car accident on the evening of November 13, 1974. Silkwood had long been a union activist at the Kerr-McGee facility and, in August 1974, had been elected to the union's bargaining committee. A month later, she and other union representatives traveled to Washington, D.C., to appear before the Atomic Energy Commission in an effort to bring to light unsafe practices at the Kerr-McGee plant. On the evening of her death, she left home to deliver documents about Kerr-McGee safety procedures to a union representative and a *New York Times* reporter. She was killed on her way to that meeting, and the documents she planned to deliver were never found.

Frederick Soddy, an English chemist who was awarded the 1921 Nobel Prize in chemistry for his research on the products of the radioactive decay of uranium and his detailed study of stable and radioactive isotopes. Soddy's research meshed nicely with that of 1922 Nobel Prize winner Francis Aston, the former concentrating on radioactive isotopes and the latter on stable isotopes.

Ernest J. Sternglass, an American physicist who worked for many years at the Westinghouse Corporation before accepting a position at the University of Pittsburgh Medical School in 1967. He researched the effects of low-level radiation on human health. As a result of his studies, Sternglass concluded that as many as 40,000 infants had died in the first two decades of the atomic era (from about 1945 to about 1965) as a result of fallout from nuclear weapons testing and radiation from other nonnatural sources. Sternglass's conclusions have long been debated but have also served as the focus of many of the protests against the development of peacetime and military applications of nuclear energy.

Fritz Strassman, a German physicist. He replaced Lise Meitner as Otto Hahn's research partner when she left Germany (because of persecution by Nazis of Jewish scientists) in 1938. Shortly after Meitner's departure, Strassman and Hahn completed experiments that demonstrated the ability of neutrons to break uranium nuclei into roughly equal halves. Strassman and Hahn were unable to accept the reality of their results, and it fell to Meitner and her nephew Otto Frisch to provide a theoretical explanation of the Hahn-Strassman experiment.

Lewis L. Strauss, the third chair of the U.S. Atomic Energy Commission (AEC). A self-made man, he was unable to attend college for financial reasons and became a traveling shoe salesman before joining the administration of President Herbert Hoover in 1917. Although he later became a wealthy investment banker, Strauss served many years in a variety of positions in the federal government. During his four-year term at the AEC, he was a fervent advocate of the development of a fusion bomb, a position that set him at odds with a number of government officials and scientists. The act for which he is probably most famous was his removal of security clearance for J. Robert Oppenheimer, one of the world's greatest nuclear experts, on the grounds that Oppenheimer, due to his reluctance to support further nuclear research, was suspected of being a Soviet agent and working to derail the U.S. fusion bomb program.

Leo Szilard, a Hungarian-American physicist. Szilard was one of the first scientists to realize the potential of nuclear energy for the development of weapons and other applications. As early as 1933, he understood the concept of a nuclear chain reaction and, only a year later, filed a patent in England for the development of such a process. For secrecy reasons, the patent was assigned not to Szilard himself but to the British Admiralty. Informed of the experiment of Otto Hahn and Fritz Strassman that resulted in the fissioning of uranium, Szilard became very concerned that Germany would develop a weapon based on the principle of nuclear fission. He approached his friend Albert Einstein, suggesting that they prepare a letter for President Franklin D. Roosevelt, outlining the concept of nuclear fission and its potential use in weapons. That letter, sent to Roosevelt on August 2, 1939, eventually led to the creation of the Manhattan Project, under which the first nuclear fission bombs were produced.

Igor Yevgenyevich Tamm, a Soviet physicist. He spent most of his professional life working on Soviet nuclear weapons and other problems of nuclear energy. Tamm worked in the Physics Institute of the USSR Academy of Sciences (FIAN) from 1934 to 1971. In 1948, he was appointed head of a special department in FIAN for the development of a fission bomb. He later was involved in research on a fusion bomb and, later still, on the development of methods for the production of energy through controlled fusion reactions. In this work, he and his colleague Andrei Sakharov designed a device known as the tokamak for the containment of fusion reactions by which energy was released. Tamm was awarded a share of the 1958 Nobel Prize in physics for his theoretical explanation of a phenomenon known as the Cerenkov effect, the production of visible light produced when radiation passes through a fluid.

Arthur Tamplin, American biophysicist. He worked for some time as a researcher on radiation issues at the Lawrence Radiation Laboratory in Livermore, California. In the late 1960s, he was asked to evaluate a study by Ernest Sternglass showing that more than 400,000 babies in the United States had died as a result of weapons testing in Nevada. Tamplin concluded that Sternglass's results may have been as much as 100 times too high and that no more than 4,000 babies had died as a result of the tests. That conclusion was still too negative as far as the AEC was concerned, and they pressed Tamplin and his colleague, John Gofman, to revise their conclusions. When Tamplin and Gofman refused to do so, the AEC cut the budgets for both researchers, refused to provide them with grant money, and made their lives so difficult that both eventually resigned from their posts at Livermore. In 1979, Tamplin and Gofman wrote a best-selling book on the safety of nuclear power plants, *Poisoned Power: The Case against Nuclear Power Plants.*

Edward Teller, a Hungarian-American physicist. He was an outspoken advocate for the development of the fusion bomb in the United States and, because of his efforts in this area, is often called the Father of the Hydrogen Bomb. Like many scientists, Teller left Europe in the early 1930s because of persecution by German Nazis and other anti-Semitic groups. After arriving in the United States, he took a position as professor of physics at George Washington University and then, in 1941, joined the Manhattan Project. In his efforts to promote the development of a fusion bomb, Teller worked actively to have J. Robert Oppenheimer's security clearance revoked, alienating a number of his colleagues. In later years, Teller also became a strong advocate for a number of military and scientific programs (including President Ronald Reagan's "Star Wars" program) that also put him at odds with many of his scientific colleagues.

George Paget Thomson, English physicist. He was awarded the 1937 Nobel Prize in physics for his research on the diffraction of electrons by crystals. In early 1940, Henry Tizard, scientific adviser to the British government, first learned about research on nuclear fission that suggested the possibility of building a nuclear weapon. He suggested the formation of a committee to investigate the significance of this work for the British military effort. Thomson, son of J. J. Thomson, the discoverer of the electron, was appointed chair of that committee. After more than a year's deliberation, Thomson's committee concluded that fission provided a viable basis for development of a nuclear weapon. The report of that committee, code-named the MAUD Committee (after a governess to Niels Bohr's children), was transmitted to the U.S. government in October 1941. The committee's conclusions eventually turned out to be of pivotal importance in convincing President Franklin D. Roosevelt to go forward with research on a fission bomb.

Joseph John Thomson (J. J.), English physicist. He was awarded the 1906 Nobel Prize in physics for his discovery of the electron. Thomson's discovery provided specific and concrete evidence, for the first time in history, that the atom is not (as John Dalton had proposed) an indivisible particle but, in fact, that it consisted of at least two parts. The negatively charged electron that he discovered constituted one of those parts, and an as-yet-to-be-discovered positive part, the other.

Grace Thorpe, daughter of Olympics legend Jim Thorpe, Native American, and antinuclear activist. She became active in the antinuclear movement in 1991 when she heard that her Native American tribe, the Sac and Fox Nation, along with 16 other tribes, had applied for a grant from the Department of Energy to host sites for the disposal of nuclear wastes. Thorpe organized a campaign to oppose the waste disposal project, a campaign that was eventually successful. She has since gone on to become an outspoken opponent of nuclear power plants.

Harry S. Truman, 33rd president of the United States. He was in office during the period when the most fundamental issues relating to the military and peacetime applications of nuclear energy were being made. He was vice president under President Franklin D. Roosevelt when Roosevelt was faced with the decision about authorizing research on a fission weapon. Truman was not involved in discussions about this question and, in fact, knew nothing about the Manhattan Project for the construction of a nuclear bomb until after he became president upon Roosevelt's death on April 12, 1945. In the debate as to whether the military or civilians should control the fate of nuclear energy, Truman tended to side with the former. However, when Congress passed the McMahon Bill in 1946 creating the Atomic Energy Commission, he signed the bill against his instincts.

Harold Urey, an American chemist. He was awarded the 1934 Nobel Prize in chemistry for his discovery of deuterium, an isotope of hydrogen with atomic mass of two (compared to "normal" hydrogen, with an atomic mass of one). Urey's expertise in the study of isotopes made him a prime candidate for work during the Manhattan Project on the separation of isotopes of uranium, one of which (uranium-235) is fissionable and another (uranium-238) is not. Ironically, Urey, a pacifist, was to see that one of the most important practical applications of deuterium was to be in construction of the fusion bomb, the most destructive weapon ever created by humans.

Paul Villard, a French physicist. He is credited with the discovery of gamma rays in 1900. While studying the physical and chemical properties of radioactive materials, Villard found that such materials emit a form of radiation similar to, but significantly more energetic than, the X-rays discovered by Roentgen in 1895. Villard's rays were, in fact, thought at first to be similar to X-rays, differing only in the amount of energy they possess. It was not until some years later that Lord Ernest Rutherford was able to describe them in sufficient detail to allow their classification as a distinct form of electromagnetic radiation, to which Rutherford gave the name gamma radiation.

Georg von Hevesy (György Hevesy), a Hungarian-Dutch-Swedish chemist. He devised the idea of using radioactive isotopes as tracers. In the late 1910s and early 1920s, von Hevesy used radioactive lead to determine the way in which that element is absorbed by plants and distributed through their body parts. Von Hevesy's experiments provided the basis for a host of applications of radioactive isotopes as tracers in industry, agriculture, research, and other fields today. For his recognition of this procedure, von Hevesy was awarded the 1943 Nobel Prize in chemistry and the 1959 Atoms for Peace Award, supported by the Ford Motor Company.

Alvin W. Weinberg, an American physicist. He has long been recognized as a leading researcher, administrator, and advocate in the field of nuclear energy. He worked with the research team at the Chicago Metallurgical Laboratory during the Manhattan Project in the development of the first plutonium-producing nuclear reactors. After World War II, Weinberg became director of research at the Oak Ridge National Laboratory (ORNL) and, in 1955, he was made director of the laboratory. In all, he spent 26 years at ORNL in one capacity or another. Among his many contributions to reactor design include the development of the first reactors for use in submarines and design of the first boiling-water reactor. Weinberg was awarded a 1980 Enrico Fermi Award for his contributions to reactor research.

Herbert George Wells (H. G.) , a British author. He is widely regarded as one of the finest science fiction writers of all time. In 1914, he wrote a novel entitled *The World Set Free*, in which he discussed a civilization that had learned how to use nuclear fission reactions for the production of energy. At first, this energy is used for productive purposes, such as atomic airplanes (of which there were 30,000 in France alone in 1943, he wrote), atomic helicopters, atomic hay trucks, atomic automobiles, atomic riveters, and atomic smelting devices. Ultimately, atomic energy is also used to make weapons that result in a world war of unbelievable devastation, where whole cities are destroyed virtually instantaneously. The wars result in surviving governments' entirely redefining the social and political institutions through which humans interact with each other. The complete text of *The World Set Free* is available online through a number of Web portals, including the Project Gutenberg Etext at http://www.gutenberg.net/dirs/etext97/twsfr10.txt.

Eugene Wigner, a Hungarian-American physicist. He made a number of contributions both to the theory of nuclear fission and to the design and development of devices used in the military and peacetime applications of nuclear energy. When nuclear fission was discovered by Otto Hahn and Fritz Strassman in 1938, Wigner had already carried out a number of theoretical calculations on the possibility of nuclear chain reactions. He became convinced that such reactions were possible and realized the significance of that fact for the development of a nuclear weapon. Concerned about the rise of fascist governments in Europe during the 1930s, Wigner decided to emigrate to the United States, where he remained for the rest of his life. During the Manhattan Project, he was assigned to the Chicago Metallurgical Laboratory, where the first operating nuclear reactor was constructed. Later he was made head of the first nuclear reactors constructed at Hanford, Washington, for the production of plutonium. One of his colleagues has called Wigner "the founder of nuclear engineering" for his work in the Hanford project.

Walter H. Zinn, Canadian-born nuclear physicist. He participated in some of the earliest experiments to determine the conditions under which a nuclear chain reaction might occur. In February 1939, well before any program for the development of nuclear energy had been launched, Zinn and Leo Szilard were attempting to determine the number of neutrons released in a fission reaction and the conditions under which that number would be sufficient to allow a chain reaction to proceed. After the Manhattan Project was created, Zinn was assigned to the Chicago Metallurgical Laboratory, where he was instrumental in the design and construction of the first nuclear reactor. It was Zinn who was responsible for removing the control rods of the reactor on the occasion of its first going into operation. At the conclusion of World War II, Zinn was appointed the first director of the Argonne National Laboratory outside Chicago. He served in that post until 1956.

CHAPTER 5

GLOSSARY

The terms used in discussions of nuclear energy are drawn from a variety of fields, including science, technology, engineering, law, and business. This chapter provides definitions for some of the most common terms and phrases used in the field of nuclear energy. An excellent and extensive glossary of terms related to nuclear energy is available on the Nuclear Regulatory Commission web site at http://www.nrc.gov/reading-rm/basic-ref/glossary.html.

agreement state A state that has signed an agreement with the Nuclear Regulatory Commission specifying the methods by which it will control the production, use, and disposal of certain quantities of nuclear wastes produced within its borders.

alpha particle The nucleus of a helium atom, consisting of two protons and two neutrons.

atomic energy Energy released during the fission or fusion of an atom; more correctly referred to as **nuclear energy.**

atomic number A number used to identify an atom, equal to the number of protons in the nucleus of that atom.

background radiation Naturally occurring radiation produced by sources outside the Earth's atmosphere (such as cosmic rays) and inside the Earth itself (such as radon released from radioactive materials).

beta particle A negatively charged electron released from a radioactive material.

binding energy The minimum amount of energy needed to separate a nucleus into the neutrons and protons of which it is composed.

boiling water reactor (BWR) A nuclear reactor in which water is used as both coolant and moderator. The water is allowed to boil in the reactor core, forming steam that is used to drive a turbine and electrical generator, thereby producing electricity.

breeder reactor A nuclear reactor that produces more nuclear fuel than it consumes.

Glossary

British thermal unit (Btu) A unit in the English system of measurement of heat; the amount of heat needed to raise the temperature of one pound of water by one degree Fahrenheit.

chain reaction A reaction in which one of the starting materials is produced in the last step of the reaction, thus permitting the reaction to become self-sustaining.

cladding A thin-walled metal tube that forms the outer jacket of a nuclear fuel rod.

compact An association of two or more states that has reached an agreement regarding the disposal of low-level nuclear wastes produced within those states.

containment structure The structure surrounding a nuclear reactor, consisting of concrete, steel, and/or other materials, designed to prevent radioactive materials produced within the reactor from escaping into the surrounding environment.

control rod A metallic unit, usually in the form of a rod, a tube, or a plate, made of a metal with a strong affinity for neutrons, such that it controls the rate of a nuclear chain reaction depending on the extent to which it is immersed into a nuclear reactor core.

control room The area in a nuclear power plant in which the instruments and personnel needed to operate a nuclear power plant are housed.

coolant A substance, such as water or liquid sodium, circulated through a nuclear reactor core for the purpose of removing heat produced during the fission of the nuclear fuel in the core.

cooling tower A large structure through which steam passes after leaving an electrical generating system, allowing the release of heat into the atmosphere.

core The central portion of a nuclear reactor that contains the nuclear fuel, moderator, and other structures needed to support the system.

criticality A state in which the number of neutrons being produced in a nuclear chain reaction is exactly equal to the number of neutrons needed to sustain the reaction.

critical mass The smallest mass of a fissionable material required to maintain a nuclear chain reaction.

decay *See* **radioactive decay.**

decommissioning The process of closing down a nuclear power plant followed by a reduction in its residual radioactivity to a level that permits the release of the property for unrestricted use.

decontamination The process by which radioactive materials are removed from a structure, area, object, or person.

deuterium An isotope of hydrogen whose nucleus consists of one proton and one neutron. Deuterium is also called heavy hydrogen. The nucleus of a deuterium atom is called a deuteron.

dose   The amount of radiation absorbed by an object or material.

fast fission   A form of fission that occurs with "fast" (high-energy) neutrons. Most fission reactions occur with slow neutrons and are known as slow fission reactions.

fissible material   In general, any material that will undergo fission. The term also has the more limited meaning of referring to isotopes that fission with slow neutrons. The three most important fissible isotopes are uranium-233, uranium-235, and plutonium-239.

fission   The process by which a large nucleus is broken apart into two isotopes of roughly equal size by a neutron.

fission products   The products formed from the fission of a nucleus. Fission products consist of isotopes of smaller size along with neutrons and other small particles.

flux   The number of particles (e.g., protons, photons, electrons) that pass through a given area within a given time.

fuel assembly   A cluster of fuel rods; also called a fuel element.

fuel cycle   The sequence of steps that occurs during the process of preparing a fissionable material for use in a nuclear power plant to its final disposal. That sequence may include processes such as mining of the metal, milling, isotopic separation and/or enrichment, fabrication of fuel elements, use in a reactor, reprocessing of the spent fuel in order to recover any remaining fissionable material, re-enrichment of the fuel material, re-fabrication into new fuel elements, and waste disposal.

fuel pellet   A small coin-shaped piece of fissionable material encased in a metallic cylinder. Fuel pellets are packed together in fuel rods, which, in turn, are collected into fuel assemblies, providing the fuel units with which nuclear reactors operate.

fuel reprocessing   The sequence of steps by means of which usable fissionable material is separated from nuclear wastes after the fuel has been removed from a nuclear reactor.

fuel rod   A long, thin, cylindrical tube that contains the fissionable material used in a nuclear reactor core.

fusion   A nuclear reaction in which two small nuclei combine to produce a single larger nucleus, with the release of very large amounts of energy.

gamma radiation   A form of electromagnetic radiation with very short wavelengths and, therefore, very large energy.

gas centrifuge   A method and a device by which the isotopes of uranium can be separated from each other. As the centrifuge rotates, the lighter isotopes of uranium (uranium-235 in particular) are thrown toward the outer edges of the centrifuge, while the heavier isotopes (uranium-238 in particular) tend to remain in the center of the device.

Glossary

gas-cooled reactor   A nuclear reactor in which some gas is used as the coolant, in contrast to the use of water, liquid sodium, or some other liquid.

gauging   Any process for measuring the thickness of an object or material.

half-life   The period of time during which one half of a radioactive isotope decays to half its original amount.

heat exchange unit   Any device or system by which heat is removed from one material and transferred to a second material.

heavy water   A form of water that contains heavy hydrogen, rather than normal hydrogen. Its chemical formula is often given as D_2O, rather than H_2O. One type of nuclear reactor uses heavy water as a coolant and is known, therefore, as a heavy water reactor.

heavy water reactor   *See* **heavy water.**

high-level waste   Spent fuel from nuclear power plants and wastes produced during the reprocessing of nuclear materials.

ionizing radiation   Any form of radiation (such as alpha, beta, gamma, or X-rays) with sufficient energy to ionize atoms or molecules.

isotope   Two or more forms of an element with the same atomic number and the same or similar chemical properties, but different atomic masses and mass numbers.

light water   A term used to describe ordinary water (H_2O) in nuclear science to distinguish it from heavy water (D_2O). Reactors in which ordinary water is used as a coolant are known as light water reactors.

light water reactor   *See* **light water.**

low-level waste   Radioactive waste materials with relatively low levels of radiation, generally produced in industrial operations, medical procedures, research activities, and nuclear fuel cycle activities distinct from those in which high-level wastes are generated.

mass defect   The amount by which the mass of an atomic nucleus is less than the total mass of the individual particles of which it is made.

mass-energy equation   The mathematical relationship developed by Albert Einstein in the early 1900s showing the equivalence between a given amount of energy or mass and expressed by the formula $E = mc^2$.

mass number   The total number of protons and neutrons found in the nucleus of an element or isotope.

megawatt hour (MWh)   A measure of power; one million watt-hours.

metastable   Having a very short half-life, that is, decaying very quickly.

mill tailings   Waste materials left behind as a result of the mining of some material. In the case of uranium mining, mill tailings are always radioactive and pose special problems in their disposal.

moderator   A material used to slow down the speed of neutrons in a nuclear chain reaction. Fission usually occurs more efficiently with slow-moving neutrons, so some system is needed to reduce their speed after being produced in a fission reaction. The two most common moderators used in nuclear reactors are water and graphite.

neutrino   A fundamental particle with no electrical charge and (probably) no mass.

neutron   A fundamental particle found in all atoms (except for that of hydrogen-1) with a mass about equal to that of a proton and carrying no electrical charge.

Nuclear Club   An informal term to describe those nations of the world that possess nuclear weapons. Members of the club currently include the United States, Russia, Great Britain, France, China, India, and Pakistan, although other nations may also possess weapons about which the rest of the world does not know.

nuclear energy   Energy produced as the result of fission, fusion, radioactive decay, or some other nuclear process.

nuclear power plant   Any facility in which electricity is generated as the result of some nuclear reaction. All such plants currently use fission reactions although the production of energy from fusion reactions is a theoretical possibility.

nuclear reactor   Any type of apparatus that makes use of a controlled nuclear chain reaction to generate energy or produce radioactive isotopes.

nuclear waste   Any material formed as a waste product at any step in the fuel cycle. Nuclear wastes include the mine tailings produced during uranium mining, waste materials generated during the enrichment of uranium, spent fuel removed from nuclear reactors, and waste materials from industrial, medical, research, and other applications of nuclear chemistry.

pellet, fuel   *See* **fuel pellet.**

pile   A now largely obsolete term used for a nuclear reactor.

plutonium   A synthetic element with atomic number 94, one of whose isotopes (plutonium-239) is fissionable.

positron   A positive electron.

positron emission tomography (PET)   A diagnostic technique in which radiation emitted by the annihilation of an electron-positron pair can be used to image body parts and body structures.

power reactor   A nuclear reactor designed to produce energy, in contrast with reactors designed for other purposes, such as the production of radioactive isotopes.

pressure vessel   A containment apparatus within which the major energy-producing system of a nuclear power plant is contained. The pressure vessel is usually built of steel, concrete, and other materials to prevent the escape of radiation into the power plant structure itself. The pressure vessel holds the fuel elements, moderator, control rods, and other supporting materials.

pressurized water reactor (PWR) A type of nuclear power plant in which water is heated to a temperature greater than its boiling point within the reactor core and then transferred to an exterior structure where it is allowed to boil, producing the steam needed to run a turbine and electrical generator.

primary system The system within a nuclear power plant that contains the coolant that circulates around and through the reactor core and that removes the heat generated by fission reactions within the core.

radiation sickness syndrome A set of medical conditions, caused by exposure to radiation, that includes, in its early stages, nausea, fatigue, vomiting, and diarrhea, followed by loss of hair, hemorrhaging, inflammation of the mouth and throat, and general loss of energy.

radioactive decay The process by which an unstable isotope releases radiation and changes into a different isotope.

radioactive isotope A natural or artificial isotope that decays spontaneously with the emission of radiation, resulting in the formation of a new and different isotope. It is also known as a radioisotope.

radiography A process by which some form of radiation, usually high-energy radiation like X- or gamma rays, is used to produce an image of an object.

reactor cooling (or coolant) system A system by which the circulation of some fluid removes heat from a reactor core and transfers it to some external site.

SAFSTOR As defined by the U.S. Nuclear Regulatory Commission, "a method of decommissioning in which the nuclear facility is placed and maintained in such condition that the nuclear facility can be safely stored and subsequently decontaminated to levels that permit release for unrestricted use."

scram The sudden shutdown of a nuclear reactor, which takes place either automatically or as the result of an operator's action, to prevent the occurrence of a malfunction within the reactor.

secondary system The steam generator, pipes, tubes, and ancillary equipment needed to extract heat from a fluid heated by the reactor core and use that heat for the production of steam.

shielding Any type of material that is able to absorb radiation and, thus, provide protection for workers in a nuclear facility.

shutdown The process by which the nuclear chain reaction in a reactor is decreased, usually by means of inserting control rods into the reactor core.

significant event Any event that creates a significant challenge to a nuclear power plant's safety system.

spent fuel Fissile material that is no longer able to maintain a nuclear chain reaction. Spent fuel is also known as depleted fuel.

Standard Technical Specifications   A list of conditions that nuclear power plants must meet in order to receive approval for operation from the Nuclear Regulatory Commission.

stochastic effects   Effects that occur randomly with health consequences that are independent of the dose received. Stochastic effects usually have no threshold value, that is, they produce their consequences at even the lowest possible level of exposure. The two main stochastic effects of nuclear radiation are cancer and genetic effects.

subcriticality   A condition in which the number of neutrons being produced in a nuclear chain reaction is less than the number produced in earlier generations of the reaction.

supercriticality   A condition in which the number of neutrons being produced in a nuclear chain reaction is greater than the number produced in earlier generations of the reaction. Supercriticality that is not brought under control can result in a reactor's going out of control.

thermonuclear weapon event   A fusion reaction. Because fusion takes place only at very high temperatures (a few million degrees), it is often referred to as a thermonuclear event.

tokamak   A doughnut-shaped machine used to constrain fusion reactions so they may be used for the safe and efficient production of energy.

tracer (radioactive)   A radioactive isotope whose location and/or movement in a system can be detected because of the radiation it emits.

transmutation (of elements)   A process by which one element or isotope is converted into a different element or isotope.

transuranic element (or waste)   An element with an atomic number greater than 92, located in a position beyond that of uranium in the periodic table. Neptunium, plutonium, americium, and californium are examples of transuranic elements. Trasuranic wastes are nuclear wastes that contain transuranic elements (primarily plutonium).

trip (reactor)   *See* **scram.**

yellowcake   A form of uranium resulting from the milling process by which uranium ore is converted to impure uranium metal.

PART II

GUIDE TO FURTHER RESEARCH

CHAPTER 6

HOW TO RESEARCH
NUCLEAR POWER ISSUES

The subject of nuclear power has been the topic of considerable discussion and debate in the United States and throughout the world for more than a half century. This chapter suggests a number of ways in which researchers can learn more about the nature of that debate, the issues that are important today, and the positions that are being argued with respect to those issues.

The way in which this book is organized provides a general outline for the steps one might follow in learning more about issues relating to nuclear power. Readers should probably begin by reading Chapter 1 of the book, which provides the scientific, technical, historical, and legal background involved in nuclear power issues. Chapters 2 through 5 provide additional information about laws and court decisions relating to nuclear power, a chronology of important events in the development of nuclear power, some important individuals who have been involved in this history, and a glossary of essential terms needed to understand the nature and applications of nuclear power.

Chapters 7 and 8 contain information as to ways in which one can continue his or her study of nuclear power. Chapter 7 provides a list of print and electronic resources that contain information and opinions on nuclear power issues. The chapter is divided in sections according to major subject areas: scientific and technical subjects, historical topics, nuclear accidents, current issues, expressions of support for and opposition to nuclear power production, and legal issues. Resources listed within each of these sections are divided into categories such as books, magazine and journal articles, pamphlets and brochures, and Internet sites. Researchers should browse through these listings and become generally familiar with the types of resources available in the chapter. They can then return later to explore specific topics in more detail.

Chapter 8 provides a list of organizations with an interest in one or another aspect of nuclear power. These organizations are divided generally into governmental agencies (international and national agencies) and nongovernmental organizations. Included in the latter grouping are a fairly wide variety of groups, such as those whose primary aim is simply to provide information about nuclear power and the issues related to it, those whose objective is to promote and encourage the development and use of nuclear power, and those whose aim it is to limit or prevent the use of nuclear power. Contact information and a general description of each organization is provided in the chapter. Researchers will find it useful to know which organizations to contact about specific subjects and the best way to obtain information from those organizations.

The remaining sections of this chapter provide suggestions about the use of print and electronic resources generally available to researchers. In addition, suggestions are offered for the somewhat specialized area of legal issues relating to nuclear power.

PRINT SOURCES

For many centuries, the primary source of information on any topic has been the library. Libraries are usually buildings that hold materials known as bibliographic resources—books, magazines, newspapers, and other periodicals and documents—as well as audiovisual materials and other sources of information. General libraries tend to vary in size from small local facilities with only a few thousand books and periodicals to mammoth collections like the Library of Congress in Washington, D.C.; the Bibliothèque Nationale de France in Paris; and the British Library in London, each of which holds millions of individual items. You can find a list of the world's major general national libraries online at "National Libraries of the World," http://www.ifla.org/VI/2/p2/national-libraries.htm.

Some libraries specialize in specific topics, ranging from business and education to environmental science and nuclear science. Examples of specialized libraries are the Monroe C. Gutman Library (education) at Harvard University; the Gulf Coast Environmental Library in Beaumont, Texas; the Jonsson Library of Government Documents at Stanford University; the Cornell Law Library in Ithaca, New York; the Plasma Physics Library at Princeton University; and the Lawrence Berkeley National Laboratory Library (nuclear science) in Berkeley, California.

Specialized libraries often have information on a topic that is not available at most general libraries. At one time, that fact was not very helpful to someone who would have to travel to Berkeley, Princeton, or some other distant location to obtain the information he or she needed. Today, most libraries, both general and specialized, have online catalogs that are entirely or partially available to anyone with access to a computer. These online catalogs often allow a researcher to locate a needed item, an item that can then be ordered through a local library by means of interlibrary loan. Further information about general and specialized libraries and about the use of interlibrary loan services can be obtained from your local school or community librarian.

LIBRARY CATALOGS

The key to accessing the vast resources of any library is the card catalog. At one time, a card catalog consisted exclusively of a collection of cards stored in wooden cabinets and arranged by title, author, and subject of all the books and other materials owned by a library. Today, most libraries also have an electronic card catalog in which that information exists in electronic files that can be accessed through computers. Some libraries have eliminated the older, physical form of their card catalog, making it possible to access their collections only through the electronic card catalog. The electronic card catalog has the advantage of being available to researchers from virtually any location, compared to the traditional physical card catalog, which can be accessed only at the library itself.

Whether one searches a library's resources by means of its physical card catalog or its electronic equivalent, that search may take any one of a number of forms. One may, for example, search for the title of a publication, by the author, by subject matter, by certain key words, by publication date, or by some other criterion. While physical catalogs tend to use only the first three of these criteria, electronic catalogs often provide researchers with a wider range of options, options that can be explored by means of advanced searches. Advanced searches allow one to search for various combinations of words and numbers, combining some terms, and requiring that others be ignored. For example, one may wish to locate books that have been written on the subject of nuclear energy only between the years 1960 and 1965, only in English, and only by an author with the last name of Black. An advanced search allows these conditions to be used in looking for items in a catalog.

Advanced searches are very helpful when an initial search produces too many results. If one looks only for the subject *nuclear energy* in a catalog, for example, one may find hundreds or thousands of entries. For example, a search for that term in the online Summit search engine, which is used by all academic libraries in the state of Oregon, returns a total of 3,380 items. Examining all the titles on that list would be very time-consuming, especially if one knows in advance that he or she is interested in only one aspect of the subject. For example, if the topic of interest were really legal issues

135

involving nuclear energy, an advanced search should be used that combines these terms, in the form "nuclear energy" and "legal issues," or some similar choice of search terms.

One of the key tools used in advanced searches is the Boolean operator. A Boolean operator is a term that tells a computer how it should treat the terms surrounding the term. The three most common Boolean operators are AND, OR, and NOT. If a computer sees two words or phrases connected by an AND, it understands that it should look only for materials in which both words or phrases occur. If the computer sees two words connected by an OR, it knows that it should search for any document that contains one word or phrase or the other, but not necessarily both. If the computer encounters two words or phrases connected by a NOT, it understands that it should ignore the specific category that follows the NOT when searching for the general category that precedes the NOT.

An important skill in searching for materials in either a library or on the Internet is to remember that documents are not always identified by a computer in the same terms in which a researcher is thinking of them. For example, a researcher may be interested in tracking down all books in a library on the subject of *nuclear power*. If he or she types that term into the library's catalog, a few hundred or a few thousand items may show up. What the researcher may not know is that many more items of interest may exist in the library's holdings. Other authors and/or librarians may use other terms to describe the same idea as expressed by *nuclear power*. Among the most common synonyms for *nuclear power* are *nuclear energy, atomic power,* and *atomic energy*. These terms do not mean exactly the same thing as *nuclear power*, but they are close enough to serve as search terms. But even these terms do not exhaust the possible range of identifiers that will produce materials of value to the researcher. Some of the other words and phrases that one might try include the following:

- structure (atomic),
- atomic,
- power production,
- nuclear reactions,
- atomic reactions,
- nuclear,
- nuclear reactor, and
- atomic reactor.

The same approach is necessary, of course, in searching for more specialized areas of nuclear energy. In looking for materials on the subject of *nuclear power plant accidents*, for example, one should search not only under that term, but also under a variety of other variations, such as:

- accidents,
- nuclear accidents,
- nuclear power plants,
- Chernobyl, and
- Three Mile Island.

One of the keys to success in finding all or most of the materials in a library on some given topic is to imagine as many different words and phrases as possible by which that topic might be identified.

SCHOLARLY ARTICLES

One area in which libraries continue to have an important advantage over Internet searching is in the use of scholarly articles. Anyone interested in nuclear energy issues will want to examine articles published in all kinds of periodicals, ranging from general interest newspapers, such as the *New York Times* and the *Washington Post*, to more specialized journals, such as *Nuclear Science and Engineering* and *Nuclear Technology*. Some periodicals make the complete content of all their issues, dating back some number of years, freely available online. Others restrict articles to members of some particular professional society (such as the American Nuclear Society) and to readers who are willing to pay for an article. Nonmembers may purchase the right to read the same articles for some fee, which varies from periodical to periodical. The general researcher will seldom be able to afford to purchase every article found on the Internet on restricted sites. His or her choice, then, is to try locating that article in a local library (usually an academic library), requesting a copy of the article through interlibrary loan, or purchasing the article online from the journal publisher.

INTERNET SOURCES

Today the resources of libraries have been greatly enhanced by the Internet and its cousin, the World Wide Web. The Internet is a vast collection of networks, each containing very large amounts of information, generally accessible from almost any kind of individual computer. The Internet was first created for use by the U.S. military in 1969 and has since expanded to include networks of every imaginable kind from every part of the world.

WEB SITES

On the Internet, data are stored in web sites, locations where information about some specific topic is to be found. That information may range from the very specific to the very general. In the case of nuclear energy, for example, one can find web sites that focus on topics as specific as one particular nuclear power plant accident (such as "4 Dead, 7 Injured in Nuclear Power Plant Accident" at http://www.japantoday.com/e/?content=news&cat=1&id=308131), or a specific material needed in nuclear power plant construction (such as "Enriched Boric Acid for Pressurized Water Reactors" at http://www.epcorp.com/NR/rdonlyres/71174EA7-B374-4934-8596-B51D105C4F30/0/w_c_01.pdf), or as general as the process of nuclear power production itself (such as "Nuclear Power: Energy for Today and Tomorrow" at http://pw1.netcom.com/~res95/energy/nuclear.html).

A good place to begin researching a topic such as nuclear energy is with a web site whose primary or exclusive focus is on this subject. Some such web sites include:

- "How Nuclear Power Works" (http://people.howstuffworks.com/nuclear-power.htm)
- "Nuclear" (http://www.ece.umr.edu/links/power/nuclearmain.htm)
- "Nuclear Energy: Do We Have Anything to Fear?" (http://wneo.org/WebQuests/TeacherWebQuests/nuclearenergy/nuclearenergy.htm)
- "Nuclear Energy Electronic Presentation Series" (http://www.nei.org/index.asp?catnum=2&catid=227)

Additional web sites on nuclear energy can be found in Chapter 7 of this book.

One benefit of general purpose web sites of this kind is that they often provide links (connections) to other web sites with information on similar or related topics. The "Nuclear" web site listed above, for example, provides links to the following web sites:

- Nuclear Reactors in USA
- Nuclear Technology Milestones (1942–1998)
- Description about Canadian Deuterium Reactor
- French Nuclear Generation Mix
- Glossary

Searching through web sites on the Internet involves three fundamental problems of which the researcher should always be aware: complexity, accuracy, and evanescence (instability). Internet web sites are related to each other in a complex, weblike fashion (hence the name World Wide *Web*), and not in a linear fashion, like the chapters in a book. When one goes looking through the Internet ("surfing the net"), one quickly heads off in dozens of different directions, often crossing and crisscrossing pathways and web sites. It is easy to get lost and forget how one arrived at a particular web site or how to get back to the beginning of a search string. One technique for keeping track of data is to print out every page that may seem to have some significance to one's research. That information may later prove to be of little or no value. But if it does turn out to be important, the researcher does not have to worry about finding it again at some time in the future. One can also mark the page containing the information as a "Favorite," using the toolbar at the top of the search engine's homepage, allowing one to return to that page if and when it is needed at a later time.

The second inherent problem with the Internet is the accuracy of web sites. Anyone can create his or her own web site with any kind of information on it. The information does not have to be true or accurate, and no outside monitor exists to tell a researcher whether the information is reliable or not. In searching for information on nuclear energy, for example, one web site could report that nuclear power plants cost 5 cents per kwh to operate, has resulted in 1,000 human deaths in the last decade, or was first discovered by Martians. There is no way of knowing which of these statements is (or are) actually factual. As a result, researchers must constantly be even more careful than they are with print materials (which are, at least, usually edited and fact checked) as to the accuracy of information found on the Internet. They can, for example, check other web sites or print materials on the same topic to verify that information given as facts is really true. They can also look carefully at web sites themselves to see if they appear to have some special argument about nuclear energy in their presentation.

Bias, and the inaccuracies it may include, is especially likely to occur in web sites on controversial issues. When people feel very strongly about a topic, they may accidentally or intentionally provide information that is incomplete, slanted, or simply wrong. Researching a topic such as nuclear power, therefore, requires a degree of caution and a willingness to double-check information more often than when conducting other forms of research.

The third problem in conducting online research is evanescence, or the tendency of web sites to disappear over time. Nothing is likely to be as frustrating for a researcher as to find a reference to a web site with what looks to be just the right information, only to receive the message "Web site not found." The web site has, usually for unknown reasons, been deleted and is no longer available on the Internet.

However, it may not really be gone forever. Sometimes the web site's address has simply been changed, a possibility that the search engine may suggest by providing possible alternative leads to the site. Or the web site may have been cached, that is, set aside in a "hidden" location in the computer's memory. The web site may then be accessed by asking the search engine to look into its "hidden memory" and pulling up the desired page. Many "dead sites" have been stored on the Internet Archive (http://www.archive.org). A word of caution: Not every site is stored here, and users need to know the exact URL to access a site. Also, most sites archived have been stripped of all their image files.

WEB INDEXES

One kind of web site—a web index—is of special interest to researchers. A web index, as the name suggests, is a web site that is organized like an outline, arranged according to subject matter and then divided into related subtopics. Web indexes are so-called because they are, in a sense, similar to the index in a book.

Probably the most popular and certainly the largest web index is Yahoo! (http://www.yahoo.com). Two other very popular web indexes are LookSmart (http://search.looksmart.com) and About.com (http://about.com). Yahoo!'s homepage provides two ways of searching a web site's content. First, one can simply type a word, phrase, or set of words into a search box. Yahoo! will then act like a search engine on its own site, looking for any and all Web pages that match the required term(s).

Second, one can select one of the major topics listed on Yahoo!'s main page and work his or her way through increasingly more specific subtopics. For example, to find web sites on nuclear power, one would go first to the major category *Science*, then to the subcategory *Energy*, and from that page to *Nuclear*. The *Nuclear* category contains about 100 web pages divided into even more specific groupings, such as *Companies, Government Agencies, Nuclear Disasters, Nuclear Engineering, Nuclear Medicine, Nuclear Power, Nuclear Waste*, and *Nuclear Weapons*. Each level of classification contains more limited and more specific topics and, hence, a more limited number of web sites to view. The more specific the topic one is researching, then, the more deeply one can go into the list of subtopics and the fewer web sites will be presented by the web index.

In some cases, a researcher may start with a major topic other than the one he or she is studying specifically. For example, information on legal aspects of nuclear power can be found not only by starting with the *Science* section on Yahoo!'s main page, but also by beginning with the *Government* section, of which *Law* is a subsection. Again, one can work one's way through sub-topics such as *Cases, History, Legal Research, Consumer*, and other categories in which issues about nuclear power might be expected to occur.

One advantage of web indexes is that they tend to be selective. That is, rather than searching for any or all web sites that feature the subject of nuclear power, for example, they try to determine the potential usefulness of such sites. In this way, researchers are less likely to have to wade through dozens of web sites with outdated, strongly biased, inaccurate, or otherwise less useful information.

One problem with some web indexes (as with may web pages and some search engines) is the prevalence of "pop-up" advertising. Pop-ups are windows that appear automatically on opening a page. Sometimes they are related to the topic, and sometimes they simply advertise general topics ("Look for Your High School Sweetheart"). Pop-ups are an important source of revenue for Internet companies and are likely to become more common in the future. They are usually a headache for researchers, for whom they are nothing other than a time-consuming distraction. One way to avoid pop-ups is to install software that recognizes and hides pop-ups as soon as they appear. Another way is to avoid certain web indexes (such as About.com) in which pop-ups tend to appear more commonly than in other sites.

SEARCH ENGINES

The second element involved in Internet research is a search engine (SE). Search engines are systems by which one can sift through the millions of web sites available online in order to find those that may contain information on some topic of interest. Search engines are amazing technological tools that accomplish this objective in a fraction of a second and then present the researcher with the names of web sites that are likely to be of interest. The difference between search engines and web indexes is not always clear, nor is it usually important that such distinctions be made.

Some of the most popular search engines now available include Google, Yahoo!, Dogpile, Ask Jeeves, AllTheWeb, HotBot, Teoma, and LookSmart. (For an exhaustive review of search engines and related topics, see SearchEngineWatch at http://searchenginewatch.com/ or SearchEngineJournal at http://www.searchenginejournal.com). By far the most widely used of these engines is Google, which claims to search out at more than 8 billion web sites.

Learning to use an SE is similar to learning other skills: The longer one does the skill, the more one learns about the process and the better one becomes at it. The easiest approach is simply to type in a word or phrase in which one is interested and press "Enter." The problem is that "easy" searching is often not efficient. A search is likely to return the names of many web sites that have little or nothing to do with the topic of interest.

For example, asking a search engine to look for *nuclear power* may produce a number of web sites that deal with specific nuclear power plants, companies that make nuclear power plant equipment, rock music groups that contain the word *nuclear* or *power* in their names, and so on.

One important key to efficient searching, then, is to be as specific as possible. If one wished to obtain information on the current status of the proposed Yucca Mountain waste disposal site, for example, it would not be very efficient to start searching just for *nuclear waste*. Even though that search would turn up web sites dealing with Yucca Mountain, it would also produce many pages having nothing to do with that specific topic. Instead, it would be better to look for some combination of terms, such as *Yucca Mountain*, *nuclear wastes*, *2004*, and *current status*. Notice that words within quotation marks will be treated by the search engine as unitary terms. A web site that uses the phrase *Yucca Mt.* or *Yucca Mtn.* exclusively will not be listed by the search engine. Nor will be a web site where the word *Yucca* only is listed or one where the name *Yucca* is misspelled.

Trial-and-error and "practice, practice, practice" are two good ways to improve one's skills in Internet searching. But formal instruction is often very helpful also. One can learn a great deal from a brother or sister, a teacher, or a friend who has experience working on the Internet. Instruction is also available on the Internet itself. For example, two web sites that provide tutorials on Internet searching are Learn the Net at http://www.learnthenet.com and Beginner's Central at http://www.northernwebs.com/bc.

Another type of search engine that is often of value is the metasearch program. Metasearch engines (also known as metacrawlers) hunt through other search engines, collecting web sites in each of those search engines they believe to be the best matches of a researcher's search terms. Among the most popular metasearch engines now available are Dogpile, Vivisimo, Kartoo, Mamma, and SurfWax.

NUCLEAR ENERGY WEB SITES

Some of the most useful information about nuclear energy issues is to be found on web sites that are wholly devoted to that subject. Those web sites may range in size from a single page to one with dozens or even hundreds of pages. The goal of such sites also range from providing a brief, general overview of the subject to covering as many aspects of the issue—such as scientific, technical, economic, social, political, aspects—as possible. An example of a general information web site on nuclear energy is the one maintained by the Energy Information Administration of the U.S. Department of Energy at http://www.eia.doe.gov/kids/non-renewable/nuclear.html.

Specialized web sites also differ from each other in the position they take on various aspects of nuclear energy. Some sites attempt to provide information in a neutral manner, offering factual information alone and allowing readers to make up their own minds about the applications of nuclear energy. An example of this kind of web site is the excellent site on Hyperphysics, maintained by C. R. Nave at Georgia State University (http://hyperphysics.phy-astr.gsu.edu/hbase/hframe.html). Other web sites exist to promote a particular point of view about nuclear energy, either attempting to encourage its support and development or opposing its use in all, or at least some, applications. Two examples of such web sites are (pronuclear power) "Frequently Asked Questions about Nuclear Energy" (http://www.formal.stanford.edu/jmc/progress/nuclear-faq.html) by John McCarthy, former Professor of Computer Science at Stanford University, and (antinuclear power) "Nuclear Power Plants" (http://www.citizen.org/cmep/energy_enviro_nuclear/nuclear_power_plants), a web site maintained by Public Citizen, a nonprofit organization. As with any resource, researchers must (1) be aware of any biases present in a web site and (2) determine the accuracy of the information from the web site, given that bias.

LEGAL RESEARCH

In one regard, much of the debate over nuclear energy is essentially a debate over legal issues. What kinds of nuclear power plants should be allowed to be built? What kinds of operational and maintenance standards should be required? What provisions should be required for the disposal of nuclear wastes? To what extent should private companies be responsible for possible accidents at power plants? Questions such as these are all answered eventually by the passage of laws at the local, state, or federal level or by the imposition of rules and regulations by executive and/or regulatory bodies. The search for legal information is of special interest and importance, therefore, for researchers interested in nuclear energy issues.

FINDING LAWS AND REGULATIONS

The best places to begin a search for laws and regulations dealing with nuclear energy are a small number of web sites that specialize in such information. These web sites are maintained both by relatively disinterested governmental agencies and by special interest groups and individuals who may or may not have some stake in supporting and opposing the development of nuclear energy. Some examples of these web sites include the following:

- Energy Law Net, operated by attorney David Blackmar at http://www. energylawnet.com/lawsregs.html. Possibly the most complete source of information on laws, regulations, and court decisions on nuclear energy currently available
- *International Journal of Nuclear Law*, at http://www.wonuc.org/law/ijnl.htm
- Internet Law Library, originally maintained by the U.S. House of Representatives legal department, http://www.lawguru.com/ilawlib
- Laws and Regulations, operated by RadWaste.org at http://www. radwaste.org/laws.htm
- Laws, Ordinances & Standards in Each Nation, a private web site, at http://www.kh.rim.or.jp/~kidax/regl/nations.html
- Radiation Related Rules, Regulations and Laws, operated by Idaho State University, http://www.physics.isu.edu/radinf/law.htm
- Ways to Access NRC's Regulations operated by the Nuclear Regulatory Commission, at http://www.nrc.gov/what-we-do/regulatory/rulemaking/access-regs.html

One of the most important web sites on nuclear laws is http://www.nrc. gov/reading-rm/doc-collections/cfr, the web site for Title 10 of the Code of Federal Regulations, which summarizes the vast majority of U.S. regulations on nuclear issues. Information on state and municipal laws and regulations dealing with nuclear energy is also available in a number of places. Perhaps the most convenient single web site is the Energy Law Net page listed above. Many state laws can also be accessed in a number of other ways. For example, the Internet Law Library provides links to laws for all U.S. states and territories, although one must then search through each state's laws to locate information on nuclear-related topics. Some web sites provide somewhat more limited information on nuclear laws. For example, the Risk Management Internet Services web site (http://www.rmis.com) summarizes state laws on radiation and radiological topics at http://www. rmis.com/db/agencyradia.php.

FINDING COURT DECISIONS

Passing laws and adopting regulations are only one step in determining the way nuclear energy can be used in the United States and other nations. Ultimately, most laws are tested in court to produce a (usually) final decision as to exactly what those laws and regulations allow and prohibit.

Locating court decisions on the Internet is somewhat similar to the process of finding laws and regulations described in the preceding section.

Web sites that contain information and laws and regulations often includes additional information on the way those laws and regulations have been interpreted by the courts.

For researchers with little background in legal matters, an excellent tutorial is available on the Internet. The tutorial is called Legal Research FAQ (frequently asked questions) and was authored by attorney Mark Eckenwiler in 1996. The tutorial provides a very readable and detailed explanation of the way court decisions are identified and how they can be located online and in print resources. The tutorial can be accessed by a number of pathways, one of which is http://www.faws.org/faws/law/research.

A number of other web sites provide suggestions for searches in specific libraries or other sources. For example, the Law Library of Congress maintains a site of this kind at http://www.loc.gov/law/public/law-faq.html. An excellent overview on using the Internet for all kinds of legal research has also been made available by Lyonette Louis-Jacques, Librarian and Lecturer in Law at University of Chicago Law School. That overview, "Legal Research Using the Internet," is available online at http://www.lib.uchicago.edu/~llou/mpoctalk.html.

CHAPTER 7

ANNOTATED BIBLIOGRAPHY

In the six decades that have passed since nuclear weapons were first used, thousands of books, magazine and journal articles, reports, pamphlets and brochures, Internet web pages, and other kinds of documents have been written on the subject of nuclear energy. Even the most thorough bibliography can include only a small fraction of these documents. The bibliography in this chapter represents a variety of references, some of which go back to the earliest days of the atomic age, but more of which are far more recent. Some sources focus on technical aspects of nuclear science, but most are written for general audience. Finally, some sources are written specifically for younger readers, but most are written for high school age and the general public.

The chapter is divided into seven major sections: (1) scientific and technical subjects, (2) historical topics, (3) nuclear accidents, (4) current issues, (5) support for nuclear power, (6) opposition to nuclear power, and (7) legal issues. A final section includes works on miscellaneous topics. The sections provided in this chapter are not mutually exclusive. Many sources could be classified into more than one category. For example, a number of books, articles, web sites, and other sources may take a stand in favor of or opposed to the use of nuclear power while providing an extensive background in the scientific and technical basis of that technology. Other resources may be designed primarily to review current issues in the field, while providing basic background information at the same time.

SCIENTIFIC AND TECHNICAL SUBJECTS

BOOKS

American Nuclear Society Standards Subcommittee on Nuclear Terminology & Units. *Glossary of Terms in Nuclear Science and Technology.* La Grange

Park, Ill.: American Nuclear Society, 1986. A glossary of essential words and phrases used in nuclear science.

Bayliss, Colin, and Keven Langley. *Nuclear Decommissioning, Waste Management, and Environmental Site Remediation*. London: Butterworth-Heinemann, 2003. A moderately technical text that discusses the problems of dismantling nuclear facilities, developing safe systems for the storage of radioactive wastes, and returning sites to useful functions, with a review of decommissioning experiences that have taken place over the previous 15 years in the United Kingdom.

Berger, John. *Nuclear Power: The Unviable Option*. Revised edition. New York: Dell, 1977. A general introduction to the topic of nuclear power that argues against the widespread use of the technology.

Berinstein, Paula. *Alternative Energy: Facts, Statistics, and Issues*. Phoenix: Oryx Press, 2001. A collection of factual and statistical information on all aspects of alternative sources of energy, nuclear power among them.

Bertel, Rosalie. *No Immediate Danger?* London: Women's Press, 1986. A public health specialist disputes industry claims that low levels of radiation pose no threat to human health.

Bethe, Hans A. *Nuclear Energy: Readings from Scientific American*. San Francisco: W. H. Freeman, 1986. A collection of articles on various aspects of nuclear energy that originally appeared in the journal *Scientific American*.

Beyer, Robert T., ed. *Foundations of Nuclear Physics*. New York: Dover Publications, 1949. A collection of 13 original articles in the field of nuclear science by researchers such as Lord Ernest Rutherford and Enrico Fermi. This book is of considerable historical interest.

Bickel, Lennard. *The Deadly Element: The Story of Uranium*. New York: Stein and Day, 1979. An account of the history of the discovery of uranium and the role it was later to play in the development of nuclear weapons and other nuclear devices.

Bodansky, David. *Nuclear Energy: Principles, Practices, and Prospects*. Woodbury, N.Y.: American Institute of Physics, 1996. Designed for readers with at least a one-year introductory course in general physics. The book discusses a variety of technical topics, including nuclear reactions, the nuclear fuel cycle, safety issues, waste disposal problems, and social and economic issues related to the use of nuclear power.

Bradley, John, ed. *Learning to Glow: A Nuclear Reader*. Tucson: University of Arizona, 2000. A collection of 24 essays that shows how ordinary individuals have been affected by developments in nuclear science, especially those exposed to radiation at Hiroshima and Nagasaki, as a result of weapons testing, and by exposure to nuclear power plant accidents.

Bromberg, Joan Lisa. *Fusion: Science, Politics, and the Invention of a New Energy Source.* Cambridge, Mass.: MIT Press, 1985. A very useful introduction to the history of fusion research from the 1950s to the early 1980s.

Bupp, Irvin C., and Jean-Claude Derian. *Light Water: How the Nuclear Dream Dissolved.* New York: Basic Books, 1978. Republished as *The Failed Promise of Nuclear Power: The Story of Light Water* (1981). The book is a discussion of the development of light water reactors, a type of nuclear reactor that is responsible for most of the electricity generated in the world, and reasons that they eventually fell out of favor in many countries, including and especially the United States.

Carbon, Max W. *Nuclear Power: Villain or Victim? Our Most Misunderstood Source of Electricity.* Madison, Wisc.: Pebble Beach Publishers, 1997. A description of the process by which nuclear energy is used to produce electricity, some of the issues involved in nuclear power plant construction, and what the advantages and disadvantages of nuclear power are over other forms of energy.

Choppin, Gregory R., and Jan Rydberg. *Nuclear Chemistry: Theory and Applications.* New York: Pergamon Press, 1980. A textbook intended for advanced science students that discusses the chemical nature and reactions of nuclear materials.

Choppin, Gregory R., Mikhail Khankhasayev, and Hans Plendl. *Chemical Separations in Nuclear Waste Management: The State of the Art and a Look to the Future.* Columbus, Ohio: Battelle Press, 2002. A survey of the mechanisms by which nuclear waste can be separated chemically into its constituent parts for storage and disposal. This is a technical presentation with some information of interest to the general reader.

Cocharan, Robert G., and Nicholas Tsoulfanidis. *The Nuclear Fuel Cycle: Analysis and Management.* La Grange Park, Ill.: American Nuclear Society, 1999. A textbook intended for college students majoring in nuclear-related subjects, covering all aspects of the series of events by which fissionable materials are mined, processed, used, and disposed of.

Cohen, Bernard Leonard. *The Nuclear Energy Option: An Alternative for the 90's.* Boulder, Colo.: Perseus Publishing, 2000. The author, a professor of physics, provides an introduction to the operation of nuclear power plants and attempts to show that they are safer for human health and the environment overall than other methods of energy production currently in use throughout the world.

Cohen, Karl, and George M. Murphy, eds. *The Theory of Isotope Separation as Applied to the Large-Scale Production of U-235.* New York: McGraw-Hill, 1951. A technical discussion of the methods that were developed for the separation of uranium isotopes during the Manhattan Project. This book is primarily of interest as an introduction to those methods.

Annotated Bibliography

Collier, John G., and Geoffrey F. Hewitt. *Introduction to Nuclear Power*, 2nd edition. London: Taylor & Francis, 2000. A textbook on nuclear power generation designed for general readers, students at the graduate and undergraduate level, and professionals in the field, describing the technology by which nuclear power is generated and attempting to assess fears as to the safety of nuclear power plants.

Crammer, J. L., and R. E. Peierls, eds. *Atomic Energy*. New York: Pelican Books, 1950. A dated publication of interest because it provides a view of the outlook for nuclear energy in the years following World War II from articles written for the general-interest magazine *Science News*.

Dean, Stephen, ed. *Prospects for Fusion Power*. New York: Elsevier Science Ltd., 1981. A collection of papers presented at two public symposia held in November 1980 and sponsored by Fusion Power Associates.

El-Hinnawai, Essam E., ed. *Nuclear Energy and the Environment*. Oxford: Pergamon Press, 1980. Volume 11 in the Environmental Sciences and Applications series, the book discusses a variety of environmental issues.

Forshier, Steven. *Essentials of Radiation Biology and Protection*. Independence, Ky.: Delmar Learning, 2001. A fairly technical treatment of radiation health issues intended for a one-semester course in the subject.

Garwin, Richard L., and Georges Charpak. *Megawatts and Megatons: The Future of Nuclear Power and Nuclear Weapons*. Chicago: University of Chicago Press, 2002. An introduction to nuclear energy intended for the general public, with an argument for strong programs of nuclear weapons control and an active promotion of nuclear power plant development.

Gephart, Roy E. *Hanford: A Conversation About Nuclear Waste and Cleanup*. Columbus, Ohio: Battelle, 2003. A geohydrologist, the author summarizes the history of the Hanford nuclear site and reviews the methods that have been used to decontaminate the area since operations were discontinued there.

Glasstone, Samuel, and Walter H. Jordan. *Nuclear Power and Its Environmental Effects*. La Grange Park, Ill.: American Nuclear Society, 1980. From the mining of uranium to the decommissioning of a nuclear power plant to the explosion of a nuclear weapon, nuclear materials exert a variety of influences on the physical and biological environment. In detail, this book discusses the impacts of these events.

Hamilton, David I. *Diagnostic Nuclear Medicine: The Physics Perspective*. New York: Springer Verlag, 2004. An upper-level textbook that presents the physical principles involved in the use of nuclear materials for diagnostic and therapeutic purposes.

Harms, A. A., K. F. Schoepf, G. H. Miley, and D. R. Kingdon. *Principles of Fusion Energy : An Introduction to Fusion Energy for Students of Science and Engineering*. Singapore: World Scientific Publishing Company, 2000. A technical text on fusion energy written for students in an introductory course in that subject. The book is accessible only to students with a substantial background in physics in general and nuclear physics in particular.

Hendee, W. R., and F. M. Edwards, eds. *Health Effects of Exposure to Low-Level Ionizing Radiation,* 2nd edition. Bristol, U.K.: Institute of Physics Publishing, 1996. An excellent overview and source book on most aspects of radiation health problems and issues.

Henderson, Harry. *Nuclear Power: A Reference Handbook.* Santa Barbara, Calif.: ABC-CLIO, 2000. A broad, general overview of nuclear power, including biographical sketches of important individuals, a chronology of events, essential facts and documents, a directory of organizations, print and nonprint resources, and a glossary of terms.

Herman, Robin. *Fusion: The Search for Endless Energy.* Cambridge, U.K.: Cambridge University Press, 1990. An overview of the process by which fusion can be used to produce useable energy, a history of efforts to do so, and some ongoing problems in the development of fusion reactors.

International Atomic Energy Agency. *Choosing the Nuclear Power Option: Factors to Be Considered.* Vienna: International Atomic Energy Agency, 1998. A booklet written by the IAEA intended for policymakers with little or no expertise in nuclear energy to assist them in determining the feasibility of nuclear energy in their own country, with information on the nuclear fuel cycle; management of radioactive waste; financial, environmental, and regulatory aspects of nuclear power; and national nuclear power policies.

———. *Direct Methods for Measuring Radionuclides in the Human Body.* Vienna: International Atomic Energy Agency, 1996. A standard reference on the methodology of radiation measurement.

———. *Nuclear Power Reactors in the World.* Vienna: International Atomic Energy Agency, 1982– . The International Atomic Energy Agency annually publishes this extensive survey of all nuclear power plants operating anywhere in the world.

Kevles, Bettyann Holtzmann. *Naked to the Bone: Medical Imaging in the Twentieth Century.* New Brunswick, N.J.: Rutgers University Press, 1997. A review of the historical development of a number of fundamental medical procedures that use various forms of radiation, including X-ray imaging, CAT scans, and positron emission tomography.

Kidd, J. S., and Renee A. Kidd. *Quarks and Sparks: The Story of Nuclear Power.* New York: Facts On File, 1999. A straightforward review of the essential aspects of nuclear power, ranging from the methods by which it is produced to the social, economic, and ethical issues raised by its use.

Lau, Foo-Sun. *A Dictionary of Nuclear Power and Waste Management: With Abbreviations and Acronyms.* New York: John Wiley & Sons, 1987. A collection of essential terms and phrases used in the field of nuclear energy.

Mackintosh, Ray, Jim Al-Khalili, and Teresa Pena. *Nucleus: A Trip Into the Heart of Matter.* Baltimore: Johns Hopkins University Press, 2001. An overview of the history of the discovery of fundamental particles, such as the proton, neutron, electron, pions, mesons, and quarks; the discovery of nuclear fission; and the applications of nuclear science in today's world.

Makhijani, Arjun, Howard Hu, and Katherine Yih, eds. *Nuclear Wastelands: A Global Guide to Nuclear Weapons Production and Its Health and Environmental Effects.* Cambridge, Mass.: MIT Press, 2000. A series of scholarly articles that reviews in detail programs for the development of nuclear weapons in every nation of the world that has the technology with an analysis of the declared and undeclared effects on the environment and human health that can be traced to these programs.

Merrick, Malcolm V. *Essentials of Nuclear Medicine,* 2nd edition. New York: Springer-Verlag, 1998. This introduction to the principles of nuclear medicine is accessible to the intelligent beginner.

Mladenovic, Milorad. *The Defining Years in Nuclear Physics, 1932–1960s.* Bristol, U.K.: Institute of Physics, 1998. Second part (chronologically) of a two-volume series on the growth and development of nuclear physics from the end of the 19th century to the end of the 1960s.

———. *History of Early Nuclear Physics, (1896–1931).* River Edge, N.J.: Word Scientific, 1992. First part (chronologically) of a two-volume series on the growth and development of nuclear physics from the end of the 19th century to the end of the 1960s.

Murray, Raymond L. *Nuclear Energy: An Introduction to the Concepts, Systems, and Applications of Nuclear Processes.* New York: Butterworth-Heinemann, 2001. A technical discussion of nuclear energy intended for students in the field.

National Research Council. *Disposition of High Level Waste: The Continuing Societal and Technical Challenges.* Washington, D.C.: National Academies Press, 2001. A report from the NRC's Board on Radioactive Waste Management that discusses topics such as scientific and technical issues, alternatives to geological disposition, improving decision making, and the importance of international cooperation.

———. *Social and Economic Aspects of Radioactive Waste Disposal: Considerations for Institutional Management.* Washington, D.C.: National Academy Press, 1984. This report addresses a variety of socioeconomic issues related to nuclear waste disposal, including site location, transportation systems, disposal schedules, regulatory systems, and effects on people living near the sites and along the transportation routes.

———. *The Waste Isolation Pilot Plant: A Potential Solution for the Disposal of Transuranic Waste.* Washington, D.C.: National Academies Press, 1996. A report by a special committee of the National Research Council appointed to study the Department of Energy's planned waste disposal site in New Mexico. The study concludes that, as long as the seals used in the plant hold, it should be possible to store wastes safely at the facility for more than 10,000 years.

Noyes, Robert. *Nuclear Waste Cleanup Technology and Opportunities.* Norwich, N.Y.: Noyes Publications, 1995. A discussion of the nature and scope of the nation's nuclear waste disposal problems, methods that have been proposed for dealing with this problem, and the current status of these technologies.

Nuclear Energy Agency. *Beneficial Uses and Production of Isotopes: 2000 Update.* Ogdensburg, N.Y.: Renouf Publishing, 2001. The latest update providing extensive statistical data and information on the production and uses of radioactive isotopes by France's nuclear energy agency.

Nuclear Energy Data. Issy-les-Moulineaux, France: Organisation for Economic Co-operation and Development, 2004. An annual publication that provides detailed statistical information about nuclear power in nations belonging to the OECD.

Nuclear Power and Health: The Implications for Health of Nuclear Power Production. WHO Regional Publications, European Series, No. 51. Copenhagen: World Health Organization, 1994. A summary and update of research conducted over the previous two decades on the health effects of radiation resulting from both normal exposure and conditions resulting from a nuclear accident during all stages of the uranium fuel cycle.

Nuclear Power Reactors in the World. Vienna: International Atomic Energy Agency, 2003. An annual publication that lists all known nuclear power plants in the world, along with data and statistics on their physical characteristics.

Owens, Anthony David. *The Economics of Uranium.* New York: Praeger, 1985. The author outlines the social, economic, scientific, technical, and political issues that surround the production, distribution, and use of uranium for military and peaceful applications in a somewhat advanced discussion of the topic.

Pochin, E., D. Beninson, H. Jammet, and A. Lafontaine. *Nuclear Power, the Environment and Man.* Vienna: International Atomic Energy Agency, 1982. An information publication that provides a general introduction to the topic of nuclear power generation.

Rahn, Frank J., Achilles G. Adamantiades, John E. Kenton, and Chaim Braun. *A Guide to Nuclear Power Technology: A Resource for Decision Making.* Malabar, Fla.: Krieger Publishing Company, 1992. A comprehensive treatment of nuclear technology, with coverage of scientific principles involved in nuclear reactors, the uranium fuel cycle, health and environmental effects, plant decommissioning, and regulatory issues.

Ramsey, Charles B., and Mohammad Modarres. *Commercial Nuclear Power: Assuring Safety for the Future.* New York: John Wiley, 1998. A professor of nuclear engineering and an employee of the U.S. Department of Energy discuss the safety issues for human health and the environment posed by nuclear energy and explain the systems used to provide protection in both instances.

Sarkisov, A. A., and L. G. LeSage. *Remaining Issues in the Decommissioning of Nuclear Powered Vessels: Including Issues Related to the Environmental Remediation of the Supporting Infrastructure.* Dordrecht, Netherlands: Kluwer, 2003. A collection of 43 technical articles in the NATO Science Series, written primarily by Russian scientists, dealing with many aspects of the decommissioning of nuclear power plants, including ecological issues, radiation and safety issues, disposal of spent nuclear fuels, decommissioning of submarines, and problems of international cooperation.

Scheider, Walter. *A Serious but Not Ponderous Book about Nuclear Energy.* Manchester, U.K.: Cavendish Press, 2001. A book developed from a five-week course taught by the author to average high school seniors outlining the fundamental idea of nuclear science.

Schull, William J. *Effects of Atomic Radiation: A Half-Century of Studies from Hiroshima and Nagasaki.* New York: Wiley-Liss, 1995. Probably the most complete review of the effects of radiation produced by the first two atomic bombs dropped on Japan in 1945, written by a geneticist from the University of Texas at Houston who joined the Atomic Bomb Casualty Commission in Japan soon after it was formed to study the effect of the bombs on survivors in the two cities.

Slater, Robert, ed. *Radioisotopes in Biology: A Practical Approach,* 2nd edition. Oxford, U.K.: Oxford University Press, 2002. A group of papers that describe the use of radioactive isotopes in a wide range of applications in the biological sciences, from the undergraduate to professional level.

Stacey, Weston M. *Nuclear Reactor Physics.* New York: Wiley Interscience, 2001. A textbook in nuclear reactor physics intended for students of the subject and of interest to only the most serious readers in the general public.

Turner, James E. *Atoms, Radiation, and Radiation Protection.* London: Elsevier Ltd., 1986. A textbook on the health effects of radiation intended primarily for professionals in the field but also containing interesting information for the general reader.

U.S. Department of Energy. *Atomic Power in Space: A History.* Washington, D.C.: U.S. Department of Energy, 1987. An extensive review of the variety of ways in which federal agencies have attempted to use nuclear materials in the nation's space program.

———. *DOE Fundamentals: Nuclear Physics and Reactor Theory,* 2 vols. Washington, D.C.: U.S. Department of Energy, 1993. A technical handbook intended for engineers and technicians involved in the operation of nuclear power plants, with some sections having information of interest to and understandable by the general reader. The book is also available online at http://www.eh.doe.gov/techstds/standard/hdbk1019/h1019v1.pdf and http://www.eh.doe.gov/techstds/standard/hdbk1019/h1019v2.pdf.

Wagner, Henry N., and Linda E. Ketchum. *Living with Radiation: The Risk, the Promise*. Baltimore: Johns Hopkins University Press, 1989. A consideration of the health risks posed by radiation from a variety of sources, including nuclear weapons and nuclear power plants, by a medical radiologist and environmental health specialist, and a medical writer.

Wagner, Robert H., Stephen M. Karesh, and James R. Halama. *Questions and Answers in Nuclear Medicine*. St. Louis: Mosby, 1999. Intended as a way of preparing for tests and examinations in nuclear medicine, this book provides an introduction to the fundamental principles of that field.

Wilson, Michael A., ed. *Textbook of Nuclear Medicine*. New York: Raven Press, 1998. A general introduction to the subject of nuclear medicine.

Wilson, P. D. *The Nuclear Fuel Cycle: From Ore to Wastes*. New York: Oxford University Press, 1996. One of the most comprehensive explanations of the uranium fuel cycle, from mining to waste storage and disposal. Special attention is given not only to the science and technology involved in the cycle but also to environmental and safety issues involved in the handling of nuclear materials.

BOOKLETS RELEASED BY THE ATOMIC ENERGY COMMISSION

In the 1960s and 1970s, the Atomic Energy Commission published a series of booklets on many aspects of nuclear power in their "Understanding the Atom" series. Although some of these booklets may be difficult to find today, they are often worth the effort. The booklets in the series are as follows (many booklets were revised after their original date of publication):

Asimov, Isaac. *World within Worlds: The Story of Nuclear Energy* (in 3 volumes), 1972. (Reissued in 2000 by University Press of the Pacific (Honolulu) under the same title.)

Asimov, Isaac, and Theodosius Dobzhansky. *The Genetic Effects of Radiation*, 1966.

A Bibliography of Basic Books on Atomic Energy, 1971.

Comar, C. L. *Fallout from Nuclear Tests*, 1963.

Corless, William R. *Computers*, 1966.

———. *Direct Conversion of Energy*, 1964.

———. *Teleoperators: Man's Machine Partners*, 1972.

Dukert, Joseph M. *Thorium and the Third Fuel*, 1970.

Faul, Henry. *Nuclear Clocks*, 1966.

Fox, Charles H. *Radioactive Wastes*, 1966.

Frigerio, Norman A., *Your Body and Radiation*, 1966.

Glasstone, Samuel. *Controlled Nuclear Fusion*, 1964.

———. *Inner Space: The Structure of the Atom*, 1972.

Gschneidner, Karl A., Jr. *Rare Earths: The Fraternal Fifteen*, 1964.

Hellman, Hal. *Atomic Particle Detection*, 1970.

———. *Spectroscopy*, 1968.

Hines, Neal O. *Atoms, Nature, and Man*, 1966.

Hogerton, John F. *Atomic Fuel.* 1963

———. *Atomic Power Safety*, 1964.

———. *Nuclear Reactors*, 1963.

Hull, E. W. Seabrook. *The Atom and Ocean*, 1968.

Hyde, Earl K. *Synthetic Transuranium Elements*, 1964.

Kernan, William J. *Accelerators*, n.d.

Kisieleski, Walter E., and Renato Baserga. *Radioisotopes and Life Processes*, 1966.

LeCompte, Robert G., and Burrell L. Wood. *Atoms at the Science Fair*, 1968.

Lyerly, Ray L., and Walter Mitchell, III. *Nuclear Power Plants*, 1967.

Martens, Frederick H., and Norman H. Jacobson, *Research Reactors*, 1965.

McIlhenny, Loyce J. *Careers in Atomic Energy*, 1962.

Mitchell, Walter, III, and Stanley E. Turner, *Breeder Reactors*, 1971.

Phelans, Earl W. *Radioisotopes in Medicine*, 1966.

Pizer, Vernon. *Preserving Food with Atomic Energy*, 1970.

Pollard, William G. *The Mystery of Matter*, 1970.

Ricciuti, Edward R. *Animals in Atomic Research*, 1967.

Singleton, Arthur L., Jr. *Sources of Nuclear Fuel*, 1968.

Swartz, Clifford E. *Microstructure of Matter*, 1965.

Urrows, Grace M. *Food Preservation by Irradiation*, 1964.

Woodburn, John H., and Frederick W. Langemann. *Whole Body Counters*, 1964.

MAGAZINES AND JOURNALS

A number of periodicals have one aspect or another of nuclear science as their exclusive or primary focus. Many of these periodicals discuss technical subjects, although some are intended for the general public. The following list includes some of the most important of these periodicals.

Annals of Nuclear Energy. An international journal reporting on developments in all aspects of reactor design and operation, as well as the uranium fuel cycle and related topics.

Atomic Energy. A technical journal that explores scientific issues of reactor design and operation.

Atoms for Peace. An international publication that focuses on issues involved in the peacetime applications of nuclear energy.

Bulletin of the Atomic Scientists. A magazine for general readers interested in nuclear issues, founded by a number of scientists and engineers who were involved in the Manhattan Project and the early development of nuclear weapons. Of all magazines and journals that discuss nuclear issues for the general public, the *Bulletin* is probably of the greatest usefulness and significance. The magazine's web site is at http://thebulletin.org.

Environment. A publication of Scientists' Institute for Public Information (SIPI) and the Helen Dwight Reid Educational Foundation intended for the general reader with a strong interest in nuclear issues.

Fusion Science and Technology. A journal of the American Nuclear Society that reports on research and development in fusion technology.

IEEE Transactions on Nuclear Science. Reports on theory, experiments, educational methods, and applications in the fields of nuclear and plasma science.

Issues in Science and Technology. A publication that discusses policy issues related to science, engineering, and medicine.

Journal of Nuclear Materials. A publication of the Institute of Nuclear Materials Management that reports on safeguards, control, accounting, nonproliferation, physical protection, packaging, transportation, and waste management of nuclear materials.

Journal of Nuclear Science and Technology. A publication of the Atomic Energy Society of Japan (in English).

Nuclear Energy. The professional journal of the British Nuclear Society.

Nuclear Engineering and Design. An international publication that reports on the engineering, design, safety, and construction of nuclear fission reactors.

Nuclear Science and Engineering. A professional journal of the American Nuclear Society that focuses on research in the area of nuclear science.

Nuclear Technology. A publication of the American Nuclear Society that discusses technical aspects of reactor design, construction, and operation.

The remaining entries in this section refer to specific articles of special interest and/or significance.

Ahearne, John F. "Radioactive Waste." *Physics Today*, vol. 50, no. 6, June 1997, pp. 22–23. An introduction to a special edition of the journal that discusses nuclear wastes. Other articles in the journal report on sources and types of nuclear wastes, technical issues, the proposed Yucca Mountain disposal site, and worldwide issues of nuclear waste disposal and control.

Annotated Bibliography

Bajaj, S. S. "Engineering Safety in Nuclear Power Plants." *Nuclear Power,* vol. 14, no. 3, 2000. An explanation of the methods by which safety is engineered into the construction of a nuclear power plant, written in a manner that can be understood by the nonspecialist.

Bebbington, William P. "The Reprocessing of Nuclear Fuels." *Scientific American.* vol. 249, December 1976, pp. 30–41. The author provides a detailed and somewhat technical introduction to one of the ongoing issues of environmental significance in the uranium fuel cycle. The information is very valuable in any current discussion of nuclear power production.

Biedscheid, J. A. "Radioactive Wastes." *Water Environment Research,* vol. 70, no. 4, June 1998, pp. 745–752. A review of the status of nuclear waste disposal programs in the United States and other parts of the world.

Charlton, J. S., J. A. Heslop, and P. Johnson. "Industrial Applications of Radioisotopes." *Physics in Technology,* vol. 6, March 1975, pp. 67–76. An extended review of some important applications of radioactive isotopes in a variety of industrial settings.

Golay, Michael W., and Neil E. Todreas. "Advanced Light-Water Reactors." *Scientific American,* vol. 262, April 1990, pp. 82–89. An excellent review of the major types of nuclear reactors—heavy water, gas-cooled, liquid metal, and light water—the last of which receives the most attention.

Hafele, Wolf. "Energy from Nuclear Power." *Scientific American,* vol. 263, no. 9, September 1990, pp. 137–144. A good summary article on nuclear power in a special *Scientific American* issue on energy from various sources.

Hoffert, Martin I., et al. "Advanced Technology Paths to Global Climate Stability: Energy for a Greenhouse Planet." *Science,* vol. 298, issue 5595, November 1, 2002, pp. 981–987. A detailed and technical consideration of the role that nuclear power will be able to play during the next few decades in affecting climate change patterns on Earth.

Hollister, Charles D., and Steven Nadis. "Burial of Radioactive Waste under the Seabed." *Scientific American,* vol. 278, no. 1, January 1998, pp. 60–65. The idea of burying nuclear wastes deep beneath the oceans may horrify environmentalists, but such an approach may be one of the safest methods currently available for disposing of spent radioactive materials.

Ion, S. E., and D. R. Bonser. "Fuel Cycles of the Future." *Nuclear Energy,* vol. 36, no. 2, April 1997, pp. 127–130. The authors review the steps involved in a nuclear fuel cycle and the environmental and health problems associated with each step. They suggest that such problems need to be considered in a holistic way in the future and outline an approach being developed by the association with which they are affiliated, British Nuclear Fuels.

Nuclear Power

Kastenburg, William E., and Luca J. Gratton. "Hazards of Managing and Disposing of Nuclear Waste." *Physics Today*, vol. 50, no. 6, June 1997, pp. 41–46. An excellent overview of the threats posed by nuclear wastes and the fundamental problems faced in disposing of these wastes adequately.

Kazimi, Mujid S. "Thorium Fuel for Nuclear Energy." *American Scientist*, vol. 91, no. 5, September–October 2003, pp. 408–413. Although thorium is not fissionable itself, it is a "fertile" element that can be converted into a nuclear fuel. Since it is far more abundant than uranium, it has long been considered as a possible substitute for that element in nuclear reactors, although the technology for achieving that objective is not yet economically feasible.

Kula, E. "Health Cost of a Nuclear Waste Repository, WIPP." *Environmental Management*, vol. 20, no. 1, January 1996, pp. 81–87. The author attempts to assess the health risks attributable to the new nuclear waste repository being built in New Mexico, with special attention to possible carcinogenic and genetic effects.

Lake, James A., Ralph G. Bennett, and John F. Kotek. "Next-Generation Nuclear Power." *Scientific American.* vol. 286, no. 1, January 2002, pp. 6–10ff. The recent history of nuclear power has not been encouraging, but it may provide the best hope for meeting the United States's future energy needs without contributing to the problem of climate change.

Marbach, G., and I. Cook. "Safety and Environment Aspects of a Fusion Power Reactor," *Fusion Engineering and Design*, vol. 46, no. 2–4, November 1999, pp. 243–254. A technical analysis of the environmental impact and safety issues related to the construction, development, and use of fusion power reactors. The authors conclude that, if safety and environmental problems can be solved, fusion power holds great promise for the future.

Marcus, Gail H. "Considering the Next Generation of Nuclear Power Plants." *Progress in Nuclear Energy*, vol. 37, no. 1–4, 2000, pp. 5–10. A discussion of the problems that are currently limiting the development of nuclear power, and a review of progress taking place in four areas that may help resolve these problems: new prototype reactors, improvements in current operating plants, development of advanced light water reactor technology, and revolutionary new design concepts—the so-called Generation IV reactors that are currently under development.

Rashad, S. M., and F. H. Hammad. "Nuclear Power and the Environment: Comparative Assessment of Environmental and Health Impacts of Electricity-Generating Systems." *Applied Energy*, vol. 65, no. 1, April 2000, pp. 211–229. The authors report on a statistical analysis of the number of deaths and injuries that result from power generation from a variety of sources and find that nuclear power is, overall, probably the safest of all methods, especially compared to traditional sources such as fossil fuels, as well as being the least harmful to the environment.

Sailor, William C., et al., "A Nuclear Solution to Climate Change?" *Science*, vol. 288, issue 5469, May 19, 2000, pp. 1,177–1,178. The authors discuss the problems involved in increasing the world's dependence on nuclear power and find that none of these problems is insurmountable.

Talbot, David. "The Next Nuclear Plant." *Technology Review*, vol. 105, no. 1, January 2002, pp. 54–59. The author describes and discusses a new type of nuclear power plant being developed in South Africa known as a pebble bed reactor, in which the reactor core is cooled by helium gas rather than water.

Tanaka, Yasumasa. "Nuclear and Environmental Risks: Problems of Communication." *Pacific and Asian Journal of Energy*, vol. 8, no. 1, June 1998, pp. 119–132. The author argues that governments have not done a very good job of communicating to the general public the relative health and environmental risks posed by nuclear power in comparison to power generated by traditional means. As a result, decisions about power plant construction have sometimes been based on irrational fears that have no basis in objective analysis of these relative risks.

Taubes, Gary. "Whose Nuclear Waste?" *Technology Review*, vol. 105, no. 1, January 2002, pp. 60–67. A general overview of the U.S. nuclear waste disposal problem and a review of the U.S. government's efforts to solve part of the problem with a nuclear waste repository in Nevada and resistance by Nevadans to that effort.

von Hippel, Frank N. "Plutonium and Reprocessing of Spent Nuclear Fuel." *Science*, vol. 293, issue 5539, September 28, 2001, pp. 2,397–2,398. A discussion of the May 2001 report of the National Energy Policy Development Group, chaired by Vice President Dick Cheney, with the conclusion that serious economic and environmental issues remain to be solved.

Wald, Matthew L. "Dismantling Nuclear Reactors." *Scientific American*, vol. 288, no. 3, March 2003, pp. 60–69. The decommissioning of a nuclear power plant presents a large variety of technical problems, environmental pollution being one that is not normally considered in adequate detail.

Waltar, Alan E. "Nuclear Technology's Numerous Uses." *Science and Technology*, vol. 20, no. 3, Spring 2004, pp. 48–54. The director of nuclear energy at the Pacific Northwest National Laboratory reviews a number of applications of nuclear energy in the fields of medicine, industry, agriculture, and research.

Whipple, C. G. "Can Nuclear Waste Be Stored Safely at Yucca Mountain?" *Scientific American*, vol. 274, no. 6, June 1996, pp. 72–79. A somewhat technical overview of the problem of nuclear waste generation and disposal, with a consideration of the federal government's plans to store radioactive materials in Yucca Mountain, Nevada.

Wilson, Jim. "Putting Nuclear Waste to Work." *Popular Mechanics*, vol. 175, June 1998, pp. 54–55. A method developed by electrical engineer Claudio Filippone for dealing with nuclear wastes.

PAMPHLETS AND BROCHURES

League of Women Voters. *A Nuclear Power Primer: Issues for Citizens.* Washington, D.C.: League of Women Voters Education Fund, 1982. A well-balanced and easily understood introduction to the subject of nuclear power.

U.S. Department of Energy, Office of Civilian Radioactive Waste Management. *Civilian Radioactive Waste Program,* Revision 3, February 2000. Available online at http://www.ocrwm.doe.gov/pm/pdf/pprev3.pdf. An extensive discussion of current plans for storing nuclear wastes at Yucca Mountain, Nevada, with an excellent review of the history of the nuclear waste disposal problem in the United States.

U.S. Nuclear Regulatory Commission. *Citizen's Guide to U.S. Nuclear Regulatory Commission Information.* Publication NUREG/BR-0010. Rev. 4. Washington, D.C.: Nuclear Regulatory Commission, n.d. Available online at http://www.nrc.gov/reading-rm/doc-collections/nuregs/brochures/br0010/br0010v4.pdf. A bibliographic source of information about brochures, pamphlets, reports, and other materials available from the NRC on topics such as nuclear reactors, nuclear materials, waste disposal, safety issues, enforcement, nuclear research, military topics, the history and organization of the commission, and related issues.

———. *Information Digest.* Publication NUREG-1350. Washington, D.C.: Nuclear Regulatory Commission. An annual publication that provides information on virtually every aspect of nuclear power production. One of the most complete sources of information available on all aspects of nuclear energy in the United States and other nations of the world.

REPORTS

A large number of government reports, most of them on technical subjects, are available from the Government Printing Office. These reports are listed on the GPO's web site at http://bookstore.gpo.gov/sb/sb-200.html. The GPO also provides links to a number of other governmental agencies that supply publications on various aspects of nuclear energy. See "Nuclear Power" at http://www.library.okstate.edu/govdocs/browsetopics/nuclearp.html.

Annotated Bibliography

Reports on scientific, technical, economic, social, political, and other aspects of nuclear energy are also available on a regular or irregular basis from a number of national, international, and industrial organizations. Access to these reports is available from the organization itself (see Chapter 8 for more information) or, in many cases, through online booksellers, such as Amazon.com or Barnes & Noble. Some examples of possible sources of such reports are the International Atomic Energy Agency, the International Energy Agency, the Nuclear Regulatory Commission, the U.S. Department of Energy, the U.S. State Department, the U.S. Congressional Budget Office, the U.S. General Accounting Office, the Nuclear Energy Institute, and various committees and subcommittees of the U.S. Senate and House of Representatives.

Department of Research and Isotopes. *Building a Better Future: Contributions of Nuclear Science and Technology.* Vienna: International Atomic Energy Agency, 1998. An excellent overview of the many applications of nuclear science in areas other than that of energy production.

Ferguson, Charles D., Tahseen Kazi, and Judith Perera. *Commercial Radioactive Sources: Surveying the Security Risks.* Monterrey, Calif.: Center for Non-Proliferation Studies, 2003. A study conducted to determine the potential security risks resulting from the commercial availability of radioactive sources, with some suggestions for minimizing and protecting against those risks.

Murray, Raymond, and Judith A. Powell, eds. *Understanding Radioactive Waste*, 4th edition. Columbus, Ohio: Battelle Press, 2003. Originally prepared as a report for the U.S. Department of Energy, the book has been revised and updated to include new legislation and ongoing issues of waste disposal in the United States and other countries of the world.

National Academy of Science, Committee on Principles and Operations. *One Step at a Time: The Staged Development of Geologic Repositories for High-Level Radioactive Waste.* Washington, D.C.: National Academy Press, 2004. Report of a committee appointed to study the problem of nuclear waste disposal in the United States and its recommendations for dealing with this problem.

Nuclear Energy Agency. *Nuclear Power Plant Life Management in a Changing Business World.* Ogdensburg, N.Y.: Renouf Publishing, 2001. A report of a workshop held in Washington, D.C., on this topic on June 26–27, 2000.

Nuclear Geophysics and Its Applications. Vienna: International Atomic Energy Agency, 1999. A technical report on the applications of nuclear isotopes to the study of geophysical problems.

Organisation for Economic Co-operation and Development. *Safety of the Nuclear Fuel Cycle.* Issy-les-Moulineaux, France: Organisation for Economic

Co-operation and Development, 1993. An extensive, detailed, and technical analysis of safety issues involved with the mining, processing, fabricating, using, and disposing of uranium and other fissionable materials.

INTERNET/WEB DOCUMENTS

Alsos: The Digital Library for Nuclear Issues. Available online. URL: http://alsos.wlu.edu/qsearch.asp?field=./Physics&past=3. Downloaded on February 9, 2005. A superb resource for a variety of important topics in nuclear science, including people and places important in the history of the field, major issues, military applications of nuclear science, and the science and technology of nuclear reactions. The site is especially strong in the area of biographical resources in nuclear science.

American Chemical Society. "The Living Textbook of Nuclear Chemistry." Available online. URL: http://livingtextbook.oregonstate.edu. Downloaded on February 9, 2005. This web site claims to "attempt to gather on a single web site a number of supplemental materials related to the study and practice of nuclear chemistry." The site includes articles on nuclear chemistry, suggested readings, a course in radiochemistry, an audiovisual course for training radiation workers, and a list of courses in nuclear chemistry.

American Nuclear Society. "American Nuclear Society." Available online. URL: http://www.ans.org. Downloaded on February 9, 2005. Homepage for the professional organization of nuclear power plant operators with detailed and extensive operation about the activities of the organization as well as links to other web sites that discuss nuclear issues.

———. "Nuclear Science and Technology and How It Influences Your Life." Available online. URL: http://www.aboutnuclear.org/home.cgi. Downloaded on February 9, 2005. A comprehensive web site that contains information on a broad range of topics in the area of nuclear science and technology.

Boyd, Rex. "Radioisotopes in Medicine." Available on line. URL: http://www.uic.com.au/nip26.htm. Downloaded on February 9, 2005. An overview of the uses of radioactive isotopes in diagnosis and therapy, with an extensive list of specific isotopes and the situations in which they are used.

Brain, Marshall. "How Nuclear Power Works." Available online. URL: http://people.howstuffworks.com/nuclear-power.htm. Downloaded on February 9, 2005. A general introduction to nuclear power that includes a discussion of nuclear fission, the operation of a nuclear power plant, potential problems with such plants, and a number of references.

Cantone, Marie Claire, and Augusto Giussani. "Isotopic Tracers in Biomedical Applications." Available online. URL: http://www.nupecc.org/iai2001/report/B43.pdf. Downloaded on February 9, 2005. A general introduction to the use of radioactive materials in a variety of biological and medical situations, including a history of the field and general principles involved in their use.

Center for Biological Monitoring. "RADNET: Information about Source Points of Anthropogenic Radioactivity." Available online. URL: http://www.davistownmuseum.org/cbm/Rad8.html. Downloaded on February 9, 2005. A very large site that contains information on many different sources of radiation made or induced by humans, including one section on safety issues related to nuclear power plants (Section 11).

Contemporary Physics Education Project. "The Nuclear Wall Chart." Available online. URL: http://www.lbl.gov/abc/wallchart. Downloaded on February 9, 2005. An interactive web site that allows users to access information on a wide range of topics related to nuclear physics, including subjects such as nuclear reactors, radioisotopes and their applications, and radiation.

Crump Institute for Biological Imaging. "Nuclear Medicine Mediabook." Available online. URL: http://www.crump.ucla.edu:8801/NM-Mediabook. Downloaded on February 9, 2005. An extensive collection of articles on the use of nuclear materials for diagnostic analysis of virtually all body systems, with separate sections on protocols, cases, tracers, and glossary.

Dunleavy, Mara. "Nuclear Energy," Available online. URL: http://www.yale.edu/ynhti/curriculum/units/1981/5/81.05.02.x.html. Downloaded on February 9, 2005. An instructional module on the subject of nuclear energy developed for the Yale–New Haven Teachers Institute in 1981. Outdated in some regards, this site is still a good source of information about fundamental principles of nuclear energy.

Energy Information Administration. "International Energy Outlook." Available online. URL: http://www.eia.doe.gov/oiaf/ieo. Downloaded on February 9, 2005. A very valuable source of data on all aspects of energy production and consumption for nearly every nation of the world, including information on the role of nuclear power in the world's energy equation. The site is revised and updated annually.

———. "Nuclear." Available online. URL: http://eia.doe.gov/fuelnuclear_njava.html. Downloaded on February 9, 2005. One of EIA's specialized sections on various types of fuels. This page has detailed sections on uranium enrichment, nuclear fuel, nuclear reactors, nuclear generation, radioactive waste and spent fuel, analysis, and forecasts.

———. "Nuclear Power." Available online. URL: http://www.eia.doe.gov/kids/non-renewable/nuclear.html. Downloaded on February 9, 2005. A general introduction to nuclear power designed for kids.

Entergy Corporation. "Welcome to Entergy Nuclear." Available online. URL: http://www.entergy-nuclear.com/Nuclear. Downloaded on February 9, 2005. Detailed information on many aspects of nuclear power generation from a company that operates eight nuclear power plants, including Arkansas Nuclear One, Grand Gulf Nuclear Station (Mississippi), River Bend Station (Louisiana), and Vermont Yankee.

Eriksson, Henrik. "Control the Nuclear Power Plant." Available online. URL: http://www.ida.liu.se/~her/npp/demo.html. Downloaded on February 9, 2005. An interactive program by which the user can initiate a random failure in a model nuclear power plant and then carry out the operations necessary to bring that failure under control.

Federal Emergency Management Administration. "Backgrounder: Nuclear Power Plant Emergency." Available online. URL: http://www.fema.gov/hazards/nuclear/radiolo.shtm. Downloaded on February 9, 2005. Information from the federal government about the dangers of nuclear radiation and how one should prepare for nuclear emergencies.

Federation of American Scientists. "Uranium Production," Available online. URL: http://www.fas.org/nuke/intro/nuke/uranium.htm. Downloaded on February 9, 2005. An extensive and detailed explanation of the way in which uranium is mined and its isotopes separated for use in nuclear weapons and nuclear reactors.

Florida Power and Light Company. "Nuclear Power Serves You." Available online. URL: http://www.fpl.com/about/nuclear/contents/nuclear_power_serves_you.shtml. Downloaded on February 9, 2005. An explanation of the way in which nuclear power plants work, with special attention to safety issues.

Gonyeau, Joseph. "The Virtual Nuclear Tourist: Nuclear Power Plants around the World," Available online. URL: http://www.nucleartourist.com. Downloaded on February 9, 2005. An introduction to nuclear power plants around the world that describes plant design and operating systems, significant events relating to nuclear power plants, information about the people who operate those plants, links to other nuclear power sites, and additional information on nuclear power.

Heeter, Robert. "FusEdWeb: Fusion Energy Educational Web Site," Available online. URL: http://fusedweb.pppl.gov. Downloaded on February 9, 2005. A web site that provides information on many aspects of fusion energy, ranging from its role in the production of stellar energy to the development of a controlled fusion reactor. This site is a service of the Princeton Plasma Physics Laboratory.

Informationskreis KernEnergie. "The Knowledge Database." Available online. URL: http://www.kernenergie.net/informationskreis/en/wissen/index.php?navid=10. Downloaded on February 9, 2005. A web site maintained by a consortium of German energy companies that provides an excellent introduction to a number of nuclear-related topics, including history, basic information about nuclear energy, how a nuclear power plant works, radiation exposure, the nuclear fuel cycle, and nuclear waste management.

Lawrence Berkeley National Laboratory. "The ABC's of Nuclear Science." Available online. URL: http://www.lbl.gov/abc. Downloaded on February 9, 2005. A general introduction to the basics of nuclear science, presented primarily by means of a Nuclear Wall Chart that shows the major features of the subject. The site also contains information on "the cosmic connection," experiments in nuclear science, a discussion of antimatter, a glossary, and information on safety issues.

Lipper, Ilan, and Jon Stone. "Nuclear Energy and Society." Available online. URL: http://www.umich.edu/~gs265/society/nuclear.htm. Downloaded on February 9, 2005. A brief but well-written general introduction to nuclear power issues designed for a course at the University of Michigan.

Makhijani, Arjun. "Comparison of Fossil Fuels and Nuclear Power: A Tabular Sketch." Available online. URL: http://www.ieer.org/ensec/no-1/comffnp.html. Downloaded on February 9, 2005. A comparison of the economic, environmental, health, and other factors related to the use of fossil fuels and nuclear power as a source of energy.

National Aeronautics and Space Administration. "Nuclear Power." Available online. URL: http://www-istp.gsfc.nasa.gov/stargaze/Snuclear/htm. Downloaded on February 9, 2005. An instructional tool about the scientific and technical aspects of nuclear power as well as its practical use in the generation of electrical power.

Nave, Carl R. "Nuclear Physics." Available online. URL: http://hyperphysics.phy-astr.gsu.edu/hbase/nuccon.html#c1. Downloaded on February 9, 2005. A section of the superb web site on "Hyperphysics" developed at Georgia State University's Department of Physics and Astronomy. This site presents technical information on all aspects of nuclear energy.

Nuclear Age Peace Foundation. "Nuclear Energy." Available online. URL: http://www.nuclearfiles.org/kinuclearenergy. Downloaded on February 9, 2005. A comprehensive web site that discusses virtually every aspect of both military and peacetime applications of nuclear energy with sections on the history of nuclear energy development, nuclear weapons, nuclear power plants, and ethical issues about the use of nuclear materials. An excellent source of many original documents of importance in the history of nuclear energy.

"Nuclear Chemistry." Available online. URL: http://www.chem.duke.edu/ ~jds/cruise_chem/nuclear/nuclear.html. Downloaded on February 9, 2005. A teaching unit prepared at Duke University, it provides a general introduction to many aspects of the field of nuclear chemistry.

"Nuclear Energy," Available online. URL: http://members.iinet.net.au/ ~hydros/nuclear/nuclear_energy.htm. Downloaded on February 9, 2005. This Australian web site is of special value because of the large number of external links it provides to other sites with information about various aspects of nuclear energy.

Nuclear Energy Agency. "Nuclear Power and Climate Change." Available online. URL: http://www.nea.fr/html/ndd/climate/climate.html. Downloaded on February 9, 2005. Report by an important international organization on the possible weather and climate effects of shifting energy production to nuclear sources.

"Nuclear Energy Institute." Available online. URL: http://www.nei.org. Downloaded on February 9, 2005. A web site rich in a variety of resources relating to nuclear issues, sponsored by the Nuclear Energy Institute, one of the major organizations of nuclear power companies in the world, providing information on topics such as nuclear technology, public policy issues, safety, finances, and data and statistics.

Nuclear Industry Association. "Nuclear-Climate Friendly Energy." Available online. URL: http://www.niauk.org. Downloaded on February 9, 2005. Web site for the organization of British nuclear energy companies, with links to a number of related organizations and background information on a variety of nuclear-related topics.

"Nuclear InfoRing." Available online. URL: http://www.radwaste.org/radring. Downloaded on February 9, 2005. A membership web site that aims to offer "up-to-date material in an informative and neutral fashion" from largely noncommercial sites with the goal of educating the public on issues of nuclear weapons, nuclear power production, radiation, and related issues.

Nuclear Management Company. "Nuclear Facts." Available online. URL: http://www.nmcco.com/education/facts/facts_home.htm. Downloaded on February 9, 2005. A company that operates six nuclear power plants maintains this site, which provides information on a variety of topics related to nuclear energy, including the advantages of nuclear energy, the environment, the history of nuclear power, nuclear waste disposal, radiation facts, safety, and security issues.

"Nuclear Reactor," Wikipedia Free Encyclopedia. Available online. URL: http://en.wikipedia.org/wiki/Nuclear_power. Downloaded on February 9, 2005. An entry in the Wikipedia Free Encyclopedia web site that provides a very complete discussion of all aspects of nuclear power, including basic scientific principles, types of reactors, the nuclear fuel cycle, history of nuclear power plants, advantages and disadvantages, statistics, and a list of nuclear energy associations and organizations.

Nuclear Reactor Laboratory, College of Engineering, University of Wisconsin at Madison. "University of Wisconsin Nuclear Reactor Tour." Available online. URL: http://reactor.engr.wisc.edu/power.html. Downloaded on February 9, 2005. Descriptions of the operation of boiling water and pressurized water reactors, with detailed information on the various elements involved in each type of reactor.

Nuclear Training Center (Ljubljana, Slovenia). "Nuclear Power of the World in Figures and Graphs." Available online. URL: http://www2.ijs.si/~icjt/nukestat. Downloaded on February 9, 2005. Extensive data and statistics on nuclear power plants from every nation in the world.

"NucNet." Available online. URL: http://www.worldnuclear.org. Downloaded on February 9, 2005. A subscription service that claims to be "the first and only" worldwide news service through which individuals and organizations interested in issues of nuclear energy use can obtain up-to-date information on the status of nuclear science throughout the world.

Oklahoma State University Library. "Nuclear Power." Available online. URL: http://www.library.okstate.edu/govdocs/browsetopics/nuclearp.html. Downloaded on February 9, 2005. A web site that is of special value because it provides links to a number of other basic web sites that discuss nuclear power topics.

"One Nuclear Place." Available online. URL: http://www.1nuclearplace.com. Downloaded on February 9, 2005. Contains newspaper articles on nearly every aspect of nuclear science, including weapons, nuclear power plants, waste disposal issues, safety issues, and the uranium fuel cycle, as well as references to books, jobs, press releases, and other nuclear-related topics.

Public Citizen. "Nuclear Power Reports & Factsheets." Available online. URL: http://www.citizen.org/cmep/energy_enviro_nuclear/nuclear_power_plants/articles.cfm?ID=11341. Downloaded on February 9, 2005. A collection of reports and informational brochures prepared by one of the nation's most prominent environmental organizations.

"Radwaste.org." Available online. URL: http://www.radwaste.org. Downloaded on February 9, 2005. A comprehensive web site providing information on virtually every aspect of nuclear waste disposal issues, including processing, storage, disposal, transportation, decommissioning of nuclear power plants, laws and regulations, and related governmental agencies.

Stone, Craig. "The Language of the Nucleus." Available online. URL: http://glossary.dataenabled.com. Downloaded on February 9, 2005. The site claims to be "the world's largest nuclear glossary" with more than 84,500 terms and 113,000 definitions.

Tennessee Valley Authority. "Nuclear Energy." Available online. URL: http://www.tva.gov/power/nuclear.htm. Downloaded on February 9, 2005. A general introduction to the topic of nuclear power production, with special emphasis on plants within the TVA system that use nuclear reactors for the generation of electrical power.

Union of Concerned Scientists. "Clean Energy: Nuclear Safety." Available online. URL: http://www.ucsusa.org/clean_energy/nuclear_safety. Downloaded on February 9, 2005. A comprehensive web site dealing with all aspects of nuclear safety issues, including case studies, analysis of nuclear safety problems, background information, testimony before Congress on the issue, and letters and comments by the UCS.

Uranium Information Centre, Ltd. "Uranium Information Centre." Available online. URL: http://www.uic.com.au. Downloaded on February 9, 2005. A web site whose objective it is to provide the citizens of Australia with information about that nation's nuclear power industry, although the information provided should be of interest and value to residents of any country of the world. The site contains a very useful section entitled "Briefing Papers" that lists more than a hundred extended reports on a large variety of energy-related topics in the areas of nuclear power for electricity, radioactive wastes, the uranium fuel cycle, radiation health issues, plant safety, reactor technology, mining and the environment, climate change, nuclear power in specific nations, and non-electricity uses of nuclear energy.

U.S. Department of Energy. "Nuclear Power in Space." Available online. URL: http://www.nuc.umr.edu/nuclear_facts/spacepower/spacepower.html. Downloaded on February 9, 2005. Reprint of a DOE brochure providing an extended amount of information on the applications of nuclear power in the U.S. space program, originally published in the late 1990s and now no longer generally available in print.

U.S. Nuclear Regulatory Commission. "Map of Power Reactor Sites." Available online. URL: http://www.nrc.gov/reactors/operating/map-power-reactors.html. Downloaded on February 9, 2005. Location of nuclear power plants in the United States along with information about those plants.

———. "Students' Corner." http://www.nrc.gov/reading-rm/basic-ref/students.html. A simplified introduction to the subject of nuclear energy, with sections on glossary, basic references, teacher notes, and games.

"U.S. Nuclear Regulatory Commission." Available online. URL: http://www.nrc.gov. Downloaded on February 9, 2005. Homepage of the federal agency responsible for the regulation of the use of nuclear materials in the United States, with extensive information on nuclear reactors, waste disposal issues, and safety matters related to the use of nuclear materials.

U.S. Office of Nuclear Energy, Science and Technology. "Moving Forward: Generation IV Nuclear Energy Systems." Available online. URL:

http://gen-iv.ne.doe.gov. Downloaded on February 9, 2005. A web site devoted to information about the latest generation of nuclear power plant technology expected to come online within the next decade or so. The web site includes a road map for this technology, documents relating to its development, a summary of Generation IV International Forum at which this technology was discussed, and a bulletin board for registered participants.

Watkins, Robert. "Nuclear Physics." Available online. URL: http://fangio. magnet.fsu.edu/~vlad/pr100/100yrs/html/chap05_toc.htm. Downloaded on February 9, 2005. A collection of more than 100 letters and articles that have been published in the journal *Physical Review* on the subject of nuclear science between 1931 and 1986.

Whitlock, Jeremy. "Canadian Nuclear FAQ." Available online. URL: http://www.nuclearfaq.ca. Downloaded on February 9, 2005. Answers to some frequently asked questions about nuclear science and technology in Canada by a physicist at Atomic Energy of Canada Limited's (AECL) Chalk River Laboratories and one-time president of the Canadian Nuclear Society.

World Information Service on Energy. "WISE Uranium Project." Available online. URL: http://www.antenna.nl/wise/uranium. Downloaded on February 9, 2005. A comprehensive collection of information on all aspects of uranium, including methods of mining and fabrication, uses in industry and weapons development, environmental and health effects, and storage and disposal methods.

World Nuclear Association, "Nuclear Portal." Available online. URL: http://www.world-nuclear.org/portal. Downloaded on February 9, 2005. An extensive variety of pages providing information on topics such as climate change, decommissioning of power plants, academic institutions, electronic and print journals, nuclear power plants, nuclear research centers, energy in general, and nuclear waste management.

———. "Radioisotopes in Industry." Available online. URL: http://www. world-nuclear.org/info/inf56.htm. Posted in October 2003. A thorough and easily understood explanation of a variety of ways in which radioactive isotopes are used in industry.

HISTORICAL TOPICS

BOOKS

Ackland, Len. *Making a Real Killing: Rocky Flats and the Nuclear West.* Albuquerque: The University of New Mexico Press, 1999. A history of the creation and development of the Rocky Flats nuclear site in Colorado, where much of the earliest research on nuclear weapons was conducted.

Nuclear Power

Ackland, Len, and Steven McGuire. *Assessing the Nuclear Age: Selections from the Bulletin of the Atomic Scientists*. Chicago: University of Chicago Press, 1986. A selection of articles that originally appeared in the journal *Bulletin of the Atomic Scientists*.

Allardice, Corbin, and Edward R. Trapnell. *The Atomic Energy Commission*. New York: Praeger, 1974. A history of the AEC written by two of its former and longtime employees.

Alperovitz, Gar. *The Decision to Use the Atomic Bomb*. New York: Vintage Books, 1996. An analysis of the controversy that developed in the early 1940s with regard to the use of fission weapons in an attack on Japan.

American Nuclear Society. *Controlled Nuclear Chain Reaction: The First 50 Years*. La Grange Park, Ill.: American Nuclear Society, 1992. A celebration of the first half-century of nuclear power production in the United States with an optimistic (if not entirely realistic) view of its probable future in this country.

Aron, Joan. *Licensed to Kill: The Nuclear Regulatory Commission and the Shoreham Power Plant*. Pittsburgh: University of Pittsburgh Press, 1998. A study of the history of the ill-fated Shoreham Nuclear Power Station, in Shoreham, New York, that closed without ever becoming operational, largely as the result of ongoing controversies among federal, state, and local governments; the Nuclear Regulatory Commission; the Long Island Lighting Company; and a variety of groups opposing the project.

Atkins, Stephen E. *Historical Encyclopedia of Atomic Energy*. Westport, Conn.: Greenwood Publishing Company, 2000. More than 450 entries cover nearly every aspect of nuclear power, from biographical sketches to important legislative and administrative decisions to historical aspects of the development of both peacetime and military applications of nuclear power in almost every country of the world.

Atomic Energy Canada. *Canada Enters the Nuclear Age: A Technical History of Atomic Energy of Canada Limited*. Montreal: McGill-Queen's University, 1997. As the title suggests, this is a review of the history of atomic energy in Canada.

Aubrey, Crispin. *Meltdown: The Collapse of the Nuclear Dream*. London: Collins and Brown, 1991. A critical review of the circumstances that led to the demise of the nuclear power industry in the United Kingdom.

Babin, Ronald. *The Nuclear Power Game*. Montreal: Black Rose Books, 1985. An analysis of the rise of nuclear power in Canada, with emphasis on the simultaneous growth of a "new power elite" that places its faith in technology, as well as the concomitant development of an antinuclear movement that this trend has produced.

Balogh, Brian. *Chain Reaction: Expert Debate and Public Participation in American Commercial Nuclear Power, 1945–1975*. Cambridge, U.K.: Cambridge

University Press, 1991. An analysis of the role played by members of the scientific community after World War II in the development of U.S. policy on the peacetime uses of nuclear power, and how and why their influence declined in the 1970s.

Beaver, William. *Nuclear Power Goes On-Line: A History of Shippingport.* Westport, Conn.: Greenwood Press, 1990. An enthusiastic history of the world's first commercial nuclear power plant with a discussion of the many positive contributions of the plant to the development of nuclear power generation in this country.

Bedford, Henry. *Seabrook Station: Citizen Politics and Nuclear Power.* Amherst: University of Massachusetts Press, 1990. A review of the 17-year effort by the Public Service Company of New Hampshire to gain permission to build a nuclear power plant at Seabrook, New Hampshire, with a consideration of the involvement of the company, regulatory agencies, and opposition groups in the effort.

Bothwell, Robert. *Nucleus: The History of Atomic Energy of Canada Limited.* Toronto: University of Toronto Press, 1988. A discussion of the early development of the nuclear power industry in Canada.

Boyer, Paul. *Fallout: A Historian Reflects on America's Half-Century Encounter with Nuclear Weapons.* Columbus: Ohio State University Press, 1998. The author covers a wide array of topics related to nuclear weapons, including the original decision to use the bombs, the interface between weapons development and religious thought, and the general impact of the development of nuclear weapons on modern culture.

Brown, G. I. *Invisible Rays: A History of Radioactivity.* Phoenix Mill, U.K.: Sutton Publishing, 2002. A wide-ranging discussion of radioactivity beginning with its discovery at the end of the 19th century up to the present day, with a review of issues involving the use of nuclear weapons and the problems associated with nuclear power production.

Brown, Jerry, and Rinaldo Brutoco. *Profiles in Power: The Antinuclear Movement and the Dawn of the Solar Age.* Old Tappan, N.J.: Twayne Publishers, 1997. Biographical sketches of ten individuals who have taken a stand against the nuclear power industry and/or federal and state governments in an effort to prevent the expansion of nuclear power plants in the United States.

Buckley, Brian. *Canada's Early Nuclear Policy: Fate, Chance, and Character.* Montreal: McGill-Queen's University Press, 2001. A discussion of the development of nuclear energy policy in Canada.

Burns, Grant. *The Atomic Papers: A Citizen's Guide to Selected Books and Articles on the Bomb, the Arms Race, Nuclear Power, the Peace Movement, and Related Issues.* Metuchen, N.J.: Scarecrow Press, 1984. A bibliography of more than 1,100 books and articles written about nuclear power prior to

1984. This is an extremely valuable resource for the early history of nuclear energy in the United States.

Burton, W. R., and C. J. Hasslam. *Nuclear Power, Pollution, and Politics.* London: Routledge, 1989. An analysis of the interaction of nuclear power production with political institutions and priorities and consequent effects on the environment.

Cantelon, Philip L., Richard G. Hewlett, and Robert C. Williams. *The American Atom: A Documentary History of Nuclear Policies from the Discovery of Fission to the Present*, 2nd edition. Philadelphia: University of Pennsylvania Press, 1991. A collection of documents that discuss the history of nuclear power, the Manhattan Project, the hydrogen bomb, nuclear testing, arms control, and peacetime applications of nuclear power.

Chernus, Ira. *Eisenhower's Atoms for Peace.* College Station: Texas A&M University Press, 2002. A critical analysis of President Dwight D. Eisenhower's address to the United Nations on July 16, 1945, with a discussion of developments in nuclear science that led up to the speech and its ultimate consequences on U.S. policy regarding the use of nuclear materials.

Clarfield, Gerard H., and William M. Wiecek. *Nuclear America: Military and Civilian Nuclear Power in the United States 1940–1980.* New York: Harper & Row, 1985. An excellent history of the early years of nuclear energy, with information on the political disputes over its control and military and peacetime applications.

Cohn, Steven Mark. *Too Cheap to Meter: An Economic and Philosophical Analysis of the Nuclear Dream.* Albany: State University of New York Press, 1997. An historical account of the development of nuclear power in the United States between 1950 and 1997, followed by a summary of current issues surrounding nuclear energy, including costs, reactor design, and safety issues. The author also explores in detail the way external factors, such as energy demand and the competitiveness of alternative forms of energy, may affect the future growth or decline of nuclear power.

Dahl, Per F. *Heavy Water and the Wartime Race for Nuclear Energy.* London: Institute of Physics Publishing, 1999. An excellent history of the discovery of heavy water, the role it played in the development of the first nuclear weapons in World War II, and the political and military intrigue involved by the Allies in ensuring that the Germans were unable to obtain the heavy water they needed for their own weapons development program.

Duffy, Robert J. *Nuclear Politics in America: A History and Theory of Government Regulation.* Lawrence: University Press of Kansas, 1997. An examination of the political context of nuclear energy, especially as it relates to change in political thought and practice in general in the United States over the preceding 50 years.

172

Duncan, Francis. *Rickover and the Nuclear Navy: The Discipline of Technology.* Annapolis, Md.: United States Naval Institute, 1990. An historian in the office of the Atomic Energy Commission describes the events involved in the approval, construction, and operation of the first nuclear vessels in the U.S. Navy.

Dyke, Richard Wayne. *Mr. Atomic Energy: Congressman Chet Holifield and Atomic Energy Affairs, 1945–1974.* Westport, Conn.: Greenwood Press, 1989. A biographical work that focuses on Holifield's 31 years in the U.S. Congress, 28 of which he also served on the Joint Committee on Atomic Energy.

Eckstein, Rick. *Nuclear Power and Social Power.* Philadelphia: Temple University Press, 1997. A critical study of the construction of two power plants, the Seabrook and Shoreham facilities, in which the author suggests that the driving forces in both instances were corporate profit rather than growth for the "general good" or considerations of public welfare or economic efficiency.

El-Genk, Mohamed S. *A Critical Review of Space Nuclear Power and Propulsion 1984–1993.* Washington, D.C.: AIP Press, 1984. A technical review of a number of proposed applications of nuclear power in space programs, including its use as a fuel, safety considerations, and radiative technologies.

Finch, Ron. *Exporting Danger.* London: Black Rose Books, 1996. Although Canada has never constructed a nuclear weapon, it has produced nuclear reactors and a number of nuclear devices and materials for export to at least 25 other nations over the past quarter century. This book reviews the history of nuclear materials manufactured and sold in Canada.

Fischer, David. *History of the International Atomic Energy Agency: The First Forty Years.* Vienna: International Atomic Energy Agency. Written on the 40th anniversary of the IAEA's founding, this book reviews the development of nuclear power throughout the world and the evolution of issues involving nuclear weapons, as well as the history of the IAEA itself.

Ford, Daniel F. *The Cult of the Atom: The Secret Papers of the Atomic Energy Commission.* New York: Simon and Schuster, 1986. The author provides a behind-the-scenes look at the development of U.S. policy on nuclear energy and its successor agencies, the Nuclear Regulatory Commission, the Energy Research and Development Administration, and the Department of Energy.

Fox, Karen. *The Chain Reaction: Pioneers of Nuclear Science.* New York: Franklin Watts, 1998. A book for young adults outlining the early history of nuclear science by way of biographical sketches of important figures.

Gantz, Kenneth, ed. *Nuclear Flight: The United States Air Force Programs for Atomic Jets, Missiles, and Rockets.* New York: Duell, Sloan, and Pearce, 1960. A collection of 22 articles that originally appeared in the *Air University Quarterly Review* and that discuss the principles of nuclear-powered flight.

Gerber, Michele Stenehjem. *On the Home Front: The Cold War Legacy of the Hanford Nuclear Site*. Lincoln: University of Nebraska Press, 2002. Hanford Nuclear Site, in Hanford, Washington, was the location of nuclear weapons development, research, and nuclear waste disposal since the days of the Manhattan Project. This book outlines the environmental effects of this work over the decades that Hanford was in operation.

Gibson, James N. *Nuclear Weapons of the United States: An Illustrated History*. Atglen, Pa.: Schiffer Publishing, 1996. A collection of photographs and sketches that purports to illustrate every nuclear weapon the United States has ever developed.

Goldschmidt, Bertrand. *The Atomic Complex: A Worldwide Political History of Nuclear Energy*. La Grange Park, Ill.: American Nuclear Society, 1982. A French chemist who worked under Marie Curie and participated briefly in U.S. efforts to build an atomic bomb provides a fascinating historical account of the development of nuclear power throughout the world, with insights into the political intrigue and opposition movements associated with that history.

Gosling, F. G. *The Manhattan Project: Making the Atomic Bomb*. Oak Ridge, Tenn.: U.S. Department of Energy, 1994. An official history of the program for the development of the first fission weapons.

Gottfried, Ted. *Enrico Fermi: Pioneer of the Atomic Age*. New York: Facts On File, 1992. A biography of one of the most important figures in the early development of nuclear energy.

Gowing, Margaret. *The Atom Bomb*. London: Butterworths, 1979. The historian and archivist of the British Atomic Energy Authority provides her views on the evolution of the world's first nuclear weapons.

Graetzer, Hans G., and David L. Anderson. *The Discovery of Nuclear Fission*. New York: Van Nostrand Reinhold, 1971. An excellent review of the early history of nuclear fission that includes some important source documents from the period.

Green, Harold P., and Alan Rosenthal. *The Joint Committee on Atomic Energy: A Study in Fusion of Governmental Power*. Washington, D.C.: George Washington University, 1961. A study of a revolutionary period in American history that involved new definitions in the way in which science and the government would interact with each other.

———. *Government of the Atom: The Integration of Powers*. New York: Atherton Press, 1963. An analysis of the question as to how the powerful new force of nuclear energy was to be monitored in the United States in the aftermath of World War II and the release of the first two atomic bombs.

Grey, Vivian. *Secret of the Mysterious Rays*. New York: Basic Books, 1966. A classic collection of essays for young readers about scientists who contributed to the early development of nuclear science.

Groueff, Stephane. *Manhattan Project: The Untold Story of the Making of the Atomic Bomb.* London: Collins, 1967. A Bulgarian-born journalist provides a history of the Manhattan Project that focuses heavily on the personalities involved in the program, as well as providing a good review of the scientific and technical problems faced by those workers.

Groves, Leslie R. *Now It Can Be Told: The Story of the Manhattan Project.* Cambridge, Mass.: Da Capo Press, 1983. The director of the Manhattan Project provides his view of the evolution of the nation's effort to build the first atomic bomb, with a discussion of both the scientific and technical issues that had to be resolved, along with the social and political problems that developed during the project.

Hecht, Gabrielle. *The Radiance of France: Nuclear Power and National Identity after World War II.* Cambridge, Mass.: MIT Press, 2000. An exploration of the technological, social, and cultural factors that contributed to the commitment of the French government at the end of World War II to the development of nuclear power to a degree that has been virtually unmatched in any other nation of the world. The term *radiance* in French refers to nuclear radiation.

Hevly, Bruce William, and John M. Findlay, eds. *The Atomic West.* Seattle: University of Washington Press, 1998. A collection of papers presented at a conference sponsored by the Center for the Pacific Northwest at the University of Washington, in which the history of nuclear materials development in the western United States was considered, with analysis of the long-term environmental, social, political, and other effects on the region resulting from this history.

Hewlett, Richard G., and Francis Duncan. *History of the United States Atomic Energy Commission.* Volume 1: *The New World, 1939–1946.* University Park: Pennsylvania State University Press, 1962. A scholarly narrative that tells of the earliest days of the development of United States policies on the peaceful and military applications of nuclear energy. The book was reprinted in 1991 by the University of California Press.

———. *History of the United States Atomic Energy Commission.* Volume 2: *Atomic Shield, 1947–1952.* University Park: Pennsylvania State University Press, 1962. A continuation of the evolution of nuclear policy as expressed in the programs and activities of the Atomic Energy Commission.

Holl, Jack M. *Argonne National Laboratory, 1946–96.* Urbana: University of Illinois Press, 1997. A history of one of the nation's major national nuclear laboratories that had its origins in the "Met Labs" at the University of Chicago, where much of the basic research on nuclear weapons was carried out, through its evolution as a primary center of research on nuclear power plants, nuclear biology, and materials science.

Johnson, Leland, and Daniel Schaffer. *Oak Ridge National Laboratory: The First Fifty Years.* Knoxville: University of Tennessee Press, 1994. One of the national laboratories created during the Manhattan Project for the development of the first atomic bombs, Oak Ridge has since been assigned a number of peacetime research projects, including the development of nuclear reactors and studies of the environment.

Jungk, Robert. *Brighter than a Thousand Suns: A Personal History of the Atomic Scientists.* New York: Harcourt Brace, 1958. One of the classic books that discusses the history of the Manhattan Project, written by a participant in that project and providing a number of personal accounts of those involved. It has since been reissued by other publishers, including Harvest Books (1970) and Sagebrush Bound (1999).

Keller, Alex. *The Infancy of Atomic Physics: Hercules in His Cradle.* Oxford: Clarendon Press, 1983. An excellent, readable introduction to the early development of nuclear physics with fascinating personal stories of many of the individuals involved in that history.

Kohn, Howard. *Who Killed Karen Silkwood?* New York: Summit Books, 1981. A fascinating analysis of the life and death of a woman who worked at a plutonium processing plant operated by the Kerr-McGee company in Crescent, Oklahoma, and who was hired to obtain information about possible illegal activities being conducted at the plant. Silkwood was killed in a automobile accident on the day before she was to hand over the information she had obtained as a result of her research.

Laurence, William Leonard. *Men and Atoms: The Discovery, the Uses, and the Future of Atomic Energy.* New York: Simon and Schuster, 1959. One of the earliest efforts to write a general history of the development of nuclear energy, with a review of both its peacetime and military applications and a somewhat overly optimistic view of the future benefits that nuclear power can bring to the world.

Loeb, Paul. *Nuclear Culture: Living and Working in the World's Largest Atomic Complex.* New York: Coward, McCann, and Geoghegan, 1982. A study of the way in which workers at the Hanford Site, where much of the work on the first nuclear weapons was carried out, thought about and dealt with the issues of working on these weapons of mass destruction.

McCaffrey, David. *The Politics of Nuclear Power: A History of the Shoreham Nuclear Power Plant.* Dordrecht, the Netherlands: Kluwer Academic Publishers, 1991. The author describes the process by which state and local government, influential activists, and corporations were all involved in the spectacular failure of the Shoreham Nuclear Power Plant, whose construction was first planned in 1967.

McKay, H. A. C. *The Making of the Atomic Age.* Oxford: Oxford University Press, 1984. An account of nuclear research in various countries during

World War II by a British nuclear chemist involved in some aspects of that work.

Meehan, Richard. *The Atom and the Fault: Experts, Earthquakes, and Nuclear Power.* Cambridge, Mass.: MIT Press, 1984. A case study of the first nuclear power plant built and owned by an independent corporation, at Vallecitos, California, a plant that was built in an area of high earthquake probabilities. The book discusses the difficult issue of how decisions are to be made when experts give conflicting advice and opinions about a scientific issue.

Miller, Richard L. *Under the Cloud: The Decades of Nuclear Testing.* New York: Two-Sixty Press, 1999. A review of nuclear testing in the United States conducted primarily in the 1950s and 1960s, with descriptions of the tests themselves as well as their ultimate consequences, both in terms of health effects and political repercussions resulting from the tests.

Miner, H. Craig. *Wolf Creek Station: Kansas Gas and Electric Company in the Nuclear Era.* Columbus: Ohio State University Press, 1993. As one reviewer describes the book, the author provides a "history in microcosm of America's hapless nuclear power industry since the Eisenhower administration." This is an intriguing story of opposition to the construction of a power plant that has turned out to be one of the most successful economically in the nation.

Mogren, Eric W. *Warm Sands: Uranium Mill Tailings Policy in the Atomic West.* Albuquerque: University of New Mexico Press, 2002. A historical account of the Uranium Mill Tailings Remedial Action Project (UMTRA) that lasted from 1978 to 1998, in which governmental and private agencies removed and buried about 40 million cubic yards of low-level radioactive uranium reduction mill tailings waste that had been collected from abandoned mill sites in 11 states and four Indian reservations.

Morone, Joseph G., and Edward J. Woodhouse. *The Demise of Nuclear Energy? Lessons for Democratic Control of Technology.* New Haven, Conn.: Yale University Press, 1989. A scholarly analysis of the reasons that the development of nuclear power plants in the United States has been so unsuccessful in the long term and the lessons this history provides for the way in which nuclear power in the future and other forms of energy can be put to better use.

Moss, Norman. *Men Who Play God: The Story of the H-Bomb and How the World Came to Live with It.* New York: Harper and Row, 1968. An engaging story of the development of the world's first fusion bombs, with a review of the science and technology involved, as well as an insight into the personalities of those involved in these programs and of the men and women who organized the protest movement against the development and use of the weapon.

Neuse, Steven M. *David E. Lilienthal: The Journey of an American Liberal.* Knoxville: University of Tennessee Press, 1996. A biography of a man who devoted most of his adult life to public service and was the first chairman of the Atomic Energy Commission.

O'Neill, Dan. *The Firecracker Boys.* New York: St. Martin's Press, 1994. An interesting story of the Atomic Energy Commission's Project Chariot, a 1950s plan to create a large new harbor in Alaska by exploding nuclear weapons along the coastline.

Orlans, Harold. *Contracting for Atoms: A Study of Public Policy Issues Posed by the Atomic Energy Commission's Contracting for Research, Development, and Managerial Services.* Washington, D.C.: Brookings Institution, 1967. The creation of a major federal governmental agency to manage a large scientific endeavor—the development of peacetime applications of nuclear energy—marked a watershed for both the federal government and the scientific community. This book analyzes the complex process by which this new kind of institution developed and evolved.

Ottaviani, Jim, et al. *Fallout.* Ann Arbor, Mich.: G.T. Labs, 2001. In a graphic-book (comic-book) style, the author and illustrators recall the history of the development of nuclear weapons, especially from the perspective of Leo Szilard and J. Robert Oppenheimer.

Pendergrass, Connie Baack. *Public Power, Politics, and Technology in the Eisenhower and Kennedy Years: The Hanford Dual-Purpose Reactor Controversy, 1956–1962.* New York: Arno Press, 1975. An account of a program at Hanford for the construction of a nuclear reactor with two functions: the generation of electricity and the production of plutonium, with the technological, economic, political, and other issues involved in that effort.

Pilat, Joseph F., Robert E. Pendley, and Charles K. Ebinger, eds. *Atoms for Peace: An Analysis after 30 Years.* Boulder, Colo.: Westview Press, 1985. A collection of essays that discusses the background, implementation, successes, and failures of President Eisenhower's 1953 program for the distribution of atomic information to countries around the world for peaceful applications.

Plastino, Ben J. *Coming of Age: Idaho Falls and the Idaho National Engineering Laboratory 1949–1990.* Chelsea, Mich.: Bookcrafters, 1998. The author, a longtime newspaper reporter in Idaho Falls, presents a historical account of the development of the National Reactor Testing Station in nearby Arco, with its consequent effects not only on the national nuclear scene but also the local environment.

Quester, George H. *Nuclear Monopoly.* New Brunswick, N.J.: Transaction Publishers, 2000, The author reviews a number of possible scenarios that the United States might have pursued during the short period of time during which it had a monopoly in the world on nuclear energy—the period between 1945 and 1949—options that included a preventative war against the Soviet Union or the use of much greater political pressure on the Soviet Union and/or other nations of the world.

Rashke, Richard L. *The Killing of Karen Silkwood: The Story Behind the Kerr-McGee Plutonium Case.* New York: Houghton Mifflin, 1981. An attempt to seek out the facts involved in the unexpected death of a young woman employed at the Kerr-McGee plant in Crescent, Oklahoma, where illegal activities involving the processing and sale of plutonium were thought to have been taking place.

Reed, Mary Beth, et al. *Savannah River Site at Fifty.* Washington, D.C.: U.S. Department of Energy, 2003. Presents a comprehensive history of the DOE's Savannah River Site, one of the major research and production facilities in the United States's nuclear complex.

Rhodes, Richard. *Dark Sun: The Making of the Hydrogen Bomb.* New York: Touchstone Books, 1995. A historical account that discusses not only the scientific and technical development of the nation's first fusion weapon but also the political intrigue and personal interactions that were part of the project.

———. *The Making of the Atomic Bomb.* New York: Simon and Schuster, 1986. The author won the Pulitzer Prize, the National Book Award, and the National Book Critics Circle Award for this history of the development of the first nuclear weapon.

———. *Nuclear Renewal: Common Sense About Energy.* New York: Penguin Books, 1993. The author attempts to find out how nuclear power development has gone so wrong in the United States when it has been so successful in some other parts of the world, such as France and Japan. He concludes that some major problems in this country have been bad design, inadequate attention to safety issues, and insufficient training of personnel.

Rockwell, Theodore. *The Rickover Effect: How One Man Made a Difference.* Lincoln, Neb.: iUniverse.com, 2002. A biography of Admiral Hyman Rickover, "Father of the Nuclear Navy," with special attention to his efforts to gain approval for the construction of the first nuclear submarine during the 1950s.

Seaborg, Glenn. *Adventures in the Atomic Age: From Watts to Washington.* New York: Farrar, Straus & Giroux, 2001. A fascinating memoir and inside look at the scientists involved in the development of nuclear science by one of the great figures in that history and chair of the Atomic Energy Commission for almost a decade.

———. *A Chemist in the White House: From the Manhattan Project to the End of the Cold War.* Washington, D.C.: American Chemical Society, 1998. A personal memoir of the author's own work in nuclear science (which includes his discovery of plutonium) and in the political realm (which includes long-term service to nine different U.S. presidents and the chair of the Atomic Energy Commission).

Seaborg, Glenn, and Benjamin S. Loeb. *The Atomic Energy Commission Under Nixon: Adjusting to Troubled Times.* N.Y.: Macmillan, 1993. A former chair of the AEC and one of his assistants describes the status and activities of the commission relatively late in its history.

Simpson, John W. *Nuclear Power from Underseas to Outer Space.* La Grange Park, Ill.: American Nuclear Society, 1995. A personalized and generally positive historical view of the development of nuclear power and its applications in a variety of situations from the former president of Westinghouse Power Systems Co. and of the American Nuclear Society.

Smyth, Henry de Wolf. *Atomic Energy for Military Purposes: The Official Report on the Development of the Atomic Bomb Under the Auspices of the United States Government.* Princeton, N.J.: Princeton University Press, 1945. One of the classics in the history of nuclear energy in the United States, originally written to provide the general public with an overview of the Manhattan Project without giving away any nuclear secrets.

Stacy, Susan M. *Proving the Principle: A History of the Idaho National Engineering and Environmental Laboratory, 1949–1999.* Idaho Falls: Idaho Operations Office of the Department of Energy, 2000. An official history of the National Reactor Testing Station (the original name of the Idaho National Engineering and Environmental Laboratory).

Stever, Donald W. *Seabrook and the Nuclear Regulatory Commission: The Licensing of a Nuclear Power Plant.* Hanover, N.H.: University Press of New England, 1980. A review of the technical, social, economic, and other issues involved in the proposed development of a nuclear power plant at Seabrook, New Hampshire, including the strong and prolonged battle against that plan by organized groups, such as the Clamshell Alliance.

Stoler, Peter. *Decline and Fail: The Ailing Nuclear Power Industry.* New York: Dodd, Mead & Company, 1985. A somewhat dated but interesting analysis of the birth and development of the nuclear power industry, its decline during the late 1970s and early 1980s, and its prospects for the future.

Strickland, Donald A. *Scientists in Politics: The Atomic Scientists Movement, 1945–46.* West Lafayette, Ind.: Purdue University Studies, 1968. An analysis of a pivotal period in U.S. history during which scientists became involved in the political process for the first time in history, with profound consequences both for the government and for the scientific community.

Sylves, Richard Terry. *The Nuclear Oracles: A Political History of the General Advisory Committee of the Atomic Energy Commission, 1947–1977.* Ames: Iowa State University Press, 1987. A history of the Atomic Energy Commission, describing its organization, responsibilities, and major projects and including biographical sketches of all 55 scientists and engineers who served on the committee during its existence from 1947 to 1977.

Taylor, Raymond W., and Samuel W. Taylor. *Uranium Fever, or No Talk under $1 Million.* New York: Macmillan, 1970. A firsthand account by two brothers who were involved in the "boom" days of uranium prospecting on the Colorado Plateau in the 1950s.

Thayer, Harry. *Management of the Hanford Engineer Works in World War II: How the Corps, Dupont and the Metallurgical Laboratory Fast Tracked the Original Plutonium Works.* New York: ASCE Press, 1996. An intriguing story of the race to build and put into operation an essential plant needed in the Manhattan Project, based in part on interviews with many of the individuals responsible for successful completion of the project.

Udall, Stewart L. *The Myths of August: A Personal Exploration of Our Tragic Cold War Affair with the Atom.* New York: Pantheon Books, 1994. An intriguing report by a former secretary of the interior and antinuclear activist who provides a very different view of the circumstances that led up to the development of the first nuclear weapons.

Walker, J. Samuel. *Containing the Atom: Nuclear Regulation in a Changing Environment, 1963–1971.* Berkeley: University of California Press, 1992. The second of two books that discuss the history of the Atomic Energy Commission and its successor agencies (see also Mazuzan) and that analyze the pressures and problems faced by the AEC during its formative years as well as the growth of the nuclear power industry itself.

———. *Permissible Dose: A History of Radiation Protection in the Twentieth Century.* Berkeley: University of California Press, 2000. An exhaustive review of the development of methods for measuring radiation exposure, determining the levels that may be "safe" for humans, and the development of standards that incorporate this information.

———. *A Short History of Nuclear Regulation, 1946–1990.* Washington, D.C.: Nuclear Regulatory Commission, 1993. An abbreviated version of Walker's two earlier books, *Controlling the Atom: The Beginnings of Nuclear Regulation, 1946–1962* and *Containing the Atom: Nuclear Regulation in a Changing Environment, 1963–1971*, about the history of the Atomic Energy Commission, the Nuclear Regulatory Commission, and the Energy Research and Development Administration, with additional information about the later two agencies in the 1970s and 1980s.

———. *Three Mile Island: A Nuclear Crisis in Historical Perspective.* Berkeley: University of California Press, 2004. Described by some reviewers as "the first comprehensive scholarly account" of the Three Mile Island nuclear disaster, the book looks in detail at the ongoing debate over nuclear power plants, the events that took place on March 28 through April 1, 1979, and the short- and long-term impacts of that disaster on the nuclear power industry in particular and the nation's view of energy production in general.

Weart, Spencer R. *Nuclear Fear: A History of Images.* Cambridge, Mass.: Harvard University Press, 1988. A study of the way in which nuclear energy has come to represent such a threat to human societies. The author traces modern fears as far back as mythical belief systems and shows how both proponents and opponents of nuclear power have manipulated the symbols of nuclear energy to strengthen their cases before the general public.

MAGAZINES AND JOURNALS

Anderson, Herbert L. "The Legacy of Fermi and Szilard." *The Bulletin of the Atomic Scientists,* vol. 5, September/October 1974, pp. 40–47. Contributions of two great nuclear scientists to the early development of the field, with a review of their work in the Manhattan Project.

Badash, Lawrence. "The Discovery of Radioactivity." *Physics Today,* vol. 2, February 1996, pp. 21–26. A historical account of the discovery of radioactivity by Henri Becquerel, including a brief biographical sketch of the scientist.

Badash, Lawrence, Elizabeth Hodes, and Adolph Tiddens. "Nuclear Fission: Reaction to the Discovery in 1939." *Proceedings of the American Philosophical Society,* vol. 130, June 1986, pp. 196–231. An interesting examination of the impact that Becquerel's discovery of radioactivity and consequent related research had on the scientific world in following decades.

Bohr, Niels, and John Archibald Wheeler. "The Mechanism of Nuclear Fission." *Physical Review,* vol 56, 1939, pp. 426–450. A scholarly paper of unusual historical interest in that it attempts to provide a theoretical explanation of nuclear fission only a few months after the phenomenon was first discovered.

Herzenberg, Caroline L., and Ruth H. Howes. "Women of the Manhattan Project." *Technology Review,* vol. 93, no. 8, November/December, 1993, pp. 32–40. Relatively little has been written about the contribution of women scientists during the Manhattan Project. This article explores this neglected facet of the project.

Segrè, Emilio. "The Discovery of Nuclear Fission." *Physics Today,* vol. 42, no. 7, July 1989, pp. 38–43. One of the early pioneers of fission research provides an inside look at the experimental and theoretical successes and failures in understanding the nature of nuclear fission.

Sime, Ruth Lewin. "Lise Meitner and the Discovery of Nuclear Fission." *Scientific American,* vol. 278, no. 1, January 1998, pp. 80–85. An interesting account of the role played by Meitner in the discovery of nuclear fission along with a discussion of the reason that Meitner received less credit for her work than she was due.

Starke, Kurt. "The Detours Leading to the Discovery of Nuclear Fission." *Journal of Chemical Education,* vol. 56, no. 12, December 1979, pp. 771–775. Progress in science sometimes seems, in retrospect, as if it has occurred in a relatively direct line, one step following another. This article shows how misleading that concept is with relationship to a developing understanding of the nature of nuclear fission.

Stuewer, Roger H. "Bringing the News of Fission to America." *Physics Today,* vol. 38, no. 12, October 1985, pp. 49–52. An account of the process by which the discovery of nuclear fission found its way from Europe to the United States in the late 1930s.

INTERNET/WEB DOCUMENTS

Allardice, Corbin, and Edward R. Trapnell. "The First Nuclear Chain Reaction." Available online. URL: http://hep.uchicago.edu/cp1.html. Downloaded on February 9, 2005. Extracts from a book entitled *The First Reactor,* published by the U. S. Department of Energy in 1982, providing firsthand reports on research that led to the first controlled fission chain reaction.

American Institute of Physics. "Center for History of Physics." Available online. URL: http://www.aip.org/history. Downloaded on February 9, 2005. A rich collection of books, articles, photographs, oral histories, and other resources dealing with the development of nuclear energy.

Argonne National Laboratory—West. "Nuclear History." Available online. URL: http://www.anlw.anl.gov/anlw_history/general_history/gen_hist. html. Downloaded on February 9, 2005. A historical overview of the types of nuclear reactors that have been developed and tested at this national laboratory.

Bracchini, Miguel A. "The History and Ethics behind the Manhattan Project." Available online. URL: http://www.me.utexas.edu/~uer/manhattan. Downloaded on February 9, 2005. An excellent review of the history of the Manhattan Project, with special emphasis on the ethical questions faced by scientists when confronted with the decision to drop or not drop the first nuclear bombs on Japan.

Department of Energy, History Division, Executive Secretariat. "Institutional Origins of the U.S. Department of Energy." Available on URL: http://66.102.7.104/search?q=cache:1JJmDQgFs-sJ:www.dpi.anl.gov/dpi2/instorig/instorig1.htm+%22institutional+origins+of+doe%22&hl=en. Downloaded on February 9, 2005. An excellent history of the precursors of the Department of Energy, including the Atomic Energy Commission, the Federal Energy Administration, and the Energy Research and Development Administration.

James, Carolyn C., "The Politics of Extravagance: The Aircraft Nuclear Propulsion Project." Available online. URL: http://www.nwc.navy.mil/press/Review/2000/spring/art5-sp0.htm. Downloaded on February 9, 2005. An historical account of how the U.S. Navy became involved in a "ponderous, pricey, and ultimately pathetic effort" to build a nuclear-powered airplane.

"National Atomic Museum." Available online. http://www.atomicmuseum.com. Homepage for "the nation's only congressionally chartered museum of nuclear science and history." The museum was created in 1969 to provide a historical resource for the major events that have made up the modern atomic age. The museum is eventually to be redesignated as the National Museum of Nuclear Science and History.

"The Nuclear History Site." Available online. http://nuclearhistory.tripod.com. Downloaded on February 9, 2005. A well-researched and complete site that discusses all aspects of the history of nuclear energy, including a historical overview, basics of radiation, fundamental principles of nuclear reactors, nuclear waste issues, nuclear power statistics, historical safety records, issues in nuclear power, history of the fission bomb, and questions and answers about nuclear weapons. A particularly useful section of the web site is "Development of the Nuclear Power Industry" at http://nuclearhistory.tripod.com/power.html.

Nuclear InfoRing. "The History of Nuclear Power Safety." Available online. URL: http://users.owt.com/smsrpm/nksafe. Downloaded on February 9, 2005. A section of the Nuclear InfoRing web site (see earlier listing) that discusses specifically and in detail the safety of nuclear power plants and accidents that have occurred in such plants.

U.S. Department of Energy. Office of Environmental Management. "Nuclear Timeline." Available online. URL: http://web.em.doe.gov/timeline. Downloaded on February 9, 2005. A very complete historical record of major developments in nuclear power since the pre–World War II period, with detailed information for each decade during that period.

U.S. Nuclear Regulatory Commission. "Our History." Available online. URL: http://www.nrc.gov/who-we-are/history.html. Downloaded on February 9, 2005. A detailed and well-documented overview of the history of federal regulation of the nuclear power industry, beginning with the Atomic Energy Commission and continuing through the creation and growth of the Nuclear Regulatory Commission.

Yankee Atomic Energy Company. "Decommissioning Yankee Rowe." Available online. URL: http://www.yankee.com. Downloaded on February 9, 2005. A very useful site in that it provides a detailed history, from planning and construction to decommissioning of a nuclear power plant.

NUCLEAR ACCIDENTS

BOOKS

Bailey, C. C. *The Aftermath of Chernobyl: History's Worst Nuclear Reactor Accident.* Dubuque, Iowa: Kendall-Hunt, 1989. An analysis and discussion of the Chernobyl nuclear reactor disaster.

Cantelon, Philip, and Robert C. Williams. *Crisis Contained: The Department of Energy at Three Mile Island.* Carbondale: Southern Illinois University Press, 2nd edition, 1984. The authors use government records, interviews with men and women involved with the Three Mile Island crisis, archival materials, and other sources to analyze the ways in which the U.S. Department of Energy responded to the accident that occurred at the plant in 1979.

Caufield, Catherine. *Multiple Exposures: Chronicles of the Radiation Age.* New York: Harper & Row, 1989. A historical review of accidents that have resulted in exposure to radiation, with the circumstances surrounding those incidents and the ways in which professionals and the general public have reacted to them.

Cheney, Glenn Alan. *Journey to Chernobyl: Encounters in a Radioactive Zone.* Chicago: Academy Chicago Publishers, 1995. A college professor of writing describes his trip to the site of the nuclear disaster at Chernobyl and his interviews with survivors of that event.

Darwell, John. *Legacy: Photographs from the Chernobyl Exclusion.* Stockport, U.K.: Dewi Lewis Publishers, 2001. A collection of photographs by one of England's best-known photographers illustrating the status of the "exclusion zone" established around the destroyed Chernobyl nuclear power facility 15 years earlier.

Ebel, Robert E. *Chernobyl and Its Aftermath: A Chronology of Events.* Washington, D.C.: Center for Strategic and International Studies, 1994. An expert in the subject of energy in Russia and the former Soviet Union, the author considers the lessons of the Chernobyl disaster for economic policy and warns of the likelihood of the accident's being repeated in Russia.

Ford, Daniel F. *Meltdown.* New York: Simon and Schuster, 1982. A vivid retelling of the events that occurred at Three Mile Island in 1979.

———. *Three Mile Island: Thirty Minutes to Meltdown.* New York: Simon and Schuster, 1982. The former director of the Union of Concern Scientist's second book on the 1979 disaster at the Three Mile Island nuclear reactor.

Fradkin, Philip L. *Fallout: An American Nuclear Tragedy*. Boulder, Colo.: Johnson Books, 2004. A study of the way in which governmental agencies have been less than forthright with the American people about the dangers of nuclear power and how, when presented with the option, they have chosen to promote national security above human health.

Fritsch, Albert J., Arthur H. Purcell, and Mary Byrd Davis. *Critical Hour: Three Mile Island, the Nuclear Legacy, and National Security*. Los Angeles: Yggdrasil Institute, 2004. The authors take advantage of the 25th anniversary of the Three Mile Island disaster to review the history of nuclear power production in the United States and to consider its possible future.

Fusco, Paul, and Magdalena Caris. *Chernobyl Legacy*. Cincinnati, Ohio: Writer's Digest Books, 2002. A group of essays with vivid photographs illustrating the effects of the world's worst nuclear accident.

Gale, Robert Peter, and Thomas Hauser. *Final Warning: The Legacy of Chernobyl*. New York: Warner Books, 1988. A first-person account of the medical problems that developed in the aftermath of the Chernobyl disaster, as told by the U.S. physician who led an international medical team that visited the site shortly after the accident.

Gould, Jay M., and Benjamin A. Goldman. *Deadly Deceit: Low-Level Radiation, High-Level Cover-Up*. New York: Four Walls Eight Windows, 1990. On the basis of their statistical studies of nuclear accidents at Chernobyl; Three Mile Island; Millstone, Connecticut; and the Savannah River nuclear plant in South Carolina, the authors conclude that the risks of low-level radiation is much greater than had previously been predicted.

Gray, Mike, and Ira Rosen. *The Warning: Accident at Three Mile Island: A Nuclear Omen for the Age of Terror*. Reissued edition. New York: W. W. Norton, 2003. A highly dramatic retelling of the events that occurred during the worst nuclear disaster in U.S. history.

Hampton, Wilborn. *Meltdown: A Race Against Nuclear Disaster at Three Mile Island: A Reporter's Story*. Cambridge, Mass.: Candlewick Press, 2001. A newspaper story–like account of the 1979 nuclear disaster at Three Mile Island, with insight as to the way in which residents of the area responded to the event. The author sets the disaster in a larger context of the rise and development of military and peacetime applications of nuclear power in the United States.

Houts, Peter S., Paul D. Cleary, and Teh-Wei Hu. *The Three Mile Island Crisis: Psychological, Social, and Economic Impacts on the Surrounding Population*. University Park: Pennsylvania State University, 1988. As its title suggests, the emphasis of this work is the impact on individuals and society of the Three Mile Island disaster rather than primarily on the technical aspects of the event itself.

Marples, David R. *Chernobyl and Nuclear Power in the USSR.* Hampshire, U.K.: Palgrave-Macmillan, 1986. An analysis of the Chernobyl disaster within the context of the Soviet Union's overall nuclear energy program.

Martin, Daniel. *Three Mile Island: Prologue or Epilogue?* Cambridge, Mass.: Ballinger, 1980. An analysis of the Three Mile Island disaster, with a consideration of the events that led up to the event as well as an analysis of the impact it may have on the development of nuclear power in the United States.

May, John. *The Greenpeace Book of the Nuclear Age: The Hidden History, the Human Cost.* New York: Pantheon Books, 1989. After an introductory chapter that reviews the fundamental principles of nuclear power generation, the book focuses on a number of accidents that have taken place in both civilian and military settings.

McKeown, William. *Idaho Falls: The Untold Story of America's First Nuclear Accident.* Toronto: ECW Press, 2003. An in-depth study and analysis of the accident that occurred at the National Reactor Testing Station on January 3, 1961, making use of interviews with relatives of the dead men and workers at the station along with documents not previously examined about this event.

Medvedev, Grigori. *No Breathing Room: The Aftermath of Chernobyl.* Boulder, Colo.: Perseus Books Group, 1994. A follow-up to the author's 1992 *The Truth about Chernobyl* (see below) that is devoted more to his problems in getting that book published, due more to Russian attempts to hide the details of the Chernobyl disaster, than to the event itself.

———. *The Truth about Chernobyl.* New York: Basic Books, Reprint edition, 1992. An account by the former deputy chief engineer at the Chernobyl Nuclear Facility as to the precise sequence of events that occurred during the 1986 disaster, with a commentary on the design flaws that create a strong possibility that the accident may occur again in other plants of similar design.

Medvedev, Zhores A. *The Legacy of Chernobyl.* Reprint edition. New York: W. W. Norton, 1992. An eminent Russian biologist provides his own view of the events that occurred during the 1986 Chernobyl disaster, with an extensive discussion of the long-term effects of that accident on the human population and the living environment in the area affected by the release of radiation.

Megaw, W. J. *How Safe? Three Mile Island, Chernobyl, and Beyond.* Toronto: Stoddart, 1987. A general explanation of the method by which nuclear power is produced, the safety factors included in a nuclear power plant, major accidents relating to nuclear power plants, and the significance of these events for the development of nuclear power.

Mould, Richard F. *Chernobyl Record: The Definitive History of the Chernobyl Catastrophe.* Bristol, U.K.: Institute of Physics Publishing, 2000. An extensive, detailed, and somewhat technical account of the events that led up to the Chernobyl disaster of 1986, the accident itself, and the events that took place following the meltdown at the Chernobyl reactor.

Murray, William. *Nuclear Turnaround: Recovery from Three Mile Island and the Lessons for the Future of Nuclear Power.* Bloomington, Ind.: 1st Books Library, 2003. A recounting of the recovery of the Three Mile Island nuclear power plant after its devastating accident in 1979, with some implications for the future of the nuclear power industry as a whole in the United States.

Osif, Bonnie Anne, Anthony J. Baratta, and Thomas W. Conkling. *TMI 25 Years Later: The Three Mile Island Nuclear Power Plant Accident and Its Impact.* University Park: Pennsylvania State University Press, 2004. A look back at the 1979 disaster based on reports, photographs, and videotapes donated to the Engineering Library at Pennsylvania State University.

Petryna, Adriana. *Life Exposed: Biological Citizens after Chernobyl.* Princeton, N.J.: Princeton University Press, 2002. The author makes use of interviews at hospitals, clinics, laboratories, state institutions, and individual citizens living in the area of the Chernobyl disaster to develop an anthropological analysis of the effect of that accident on the surrounding area, in particular, and the nation of Ukraine, in general.

Pligt, Joop van der. *Nuclear Energy and the Public.* Oxford: Blackwell, 1992. Based on interviews with members of the general public after the Chernobyl and Three Mile Island nuclear accidents, the author attempts to assess interviewees' understanding of, perceptions about, and opinions concerning nuclear power and the risks it poses to humans and the environment.

Polidori, Robert. *Zones of Exclusion: Pripyat and Chernobyl.* Göttingen, Germany: Steidl Verlage Publishers, 2003. Photographer Polidori returns to the Chernobyl nuclear power facility and the nearby town of Pripyat to record the long-term effects of the world's worst nuclear disaster on the area surrounding the destroyed plant.

Read, Piers Paul. *Ablaze: The Story of the Heroes and Victims of Chernobyl.* New York: Random House, 1993. An account of the Chernobyl disaster with a heavy emphasis on the role of people directly involved in the accident and a discussion of the probable role of plant workers in the disaster itself.

Rees, Joseph V. *Hostages of Each Other: The Transformation of Nuclear Safety Since Three Mile Island.* Chicago: University of Chicago Press, 1994. An analysis of the ways in which the nuclear industry's approach to safety issues changed as a result of the accident at Three Mile Island.

Silver, L. Ray. *Fallout from Chernobyl.* St. John's, Newfoundland: Breakwater Books, 1987. A description of the 1986 disaster at Chernobyl.

Stephens, Mark. *Three Mile Island.* New York: Random House, 1980. Purports to provide the real answers for the first time as to what actually happened at the nation's worst nuclear power plant disaster in 1979.

Yaroshinskaya, Alla. *Chernobyl: The Forbidden Truth.* Translated by Julia Sallabank. Lincoln: University of Nebraska Press, 1995. The author, a journalist at the time of the Chernobyl disaster, was later elected to the Russian parliament and gained access to records about the accident that were not generally available to other researchers, allowing her to write an account of the event containing more intimate details than many other reports.

MAGAZINES AND JOURNALS

Anspaugh, Lynn R., Catlin, Robert J., and Goldman, Marvin. "The Global Impact of the Chernobyl Reactor Accident." *Science*, vol. 242, no. 4885, December 16, 1988, pp. 1,513–1,519. An analysis of the potential effects of fallout of radioactive materials on Europe and North America as the result of the 1986 Chernobyl accident.

Flavin, Christopher, and Nicholas Lenssen. "Nuclear Power Browning Out." *Bulletin of the Atomic Scientists*, vol. 52, no. 3, May/June 1996, pp. 52–56. An overview of the decline of nuclear power production in the United States and the rest of the world, largely as the result of the accidents at Three Mile Island and Chernobyl.

Freemantle, Michael. "Ten Years After Chernobyl: Consequences Are Still Emerging." *Chemical & Engineering News*, vol. 74, no. 18, April 19, 1996, pp. 18–28. A summary of the health and environment effects of the Chernobyl disaster a decade after it occurred.

Gale, R. P. "Immediate Medical Consequences of Nuclear Accidents. Lessons from Chernobyl." *JAMA*, vol. 258, no. 5, August 7, 1987, pp. 625–628. A review of the information known about medical effects of the Chernobyl disaster about a year after it had occurred.

Ginzburg, Harold M., and Eric Reis. "Consequences of the Nuclear Power Plant Accident at Chernobyl." *Public Health Reports*, vol. 106, no. 1, January/February 1991, pp. 32–40. One of the most complete discussions of the health effects produced by the Chernobyl nuclear disaster in 1986.

Lewis, Harold W. "The Safety of Fission Reactors." *Scientific American*, vol. 242, no. 3, March 1980, pp. 53–65. An excellent general overview of the safety issues involved in fission reactors and how they are built to provide protection from accidents.

Scherbak, Yuri M. "Ten Years of the Chernobyl Era." *Scientific American*, vol. 274, no. 4, April 1996, pp. 44–49. The author reviews the long-term effects of the disaster at Chernobyl, some of which are still unknown and many of which are not well understood, as is the case with the accident itself.

REPORTS

International Symposium on Severe Accidents in Nuclear Power Plants. *Severe Accidents in Nuclear Power Plants: Proceedings of an International Symposium on Severe Accidents in Nuclear Power Plants.* Vienna: International Atomic Energy Agency, 2002. Technical reports on the worst accidents that have occurred in nuclear power plants throughout the world, as of the early 21st century.

Presidential Commission on Catastrophic Nuclear Accidents. *Report to the Congress from the Presidential Commission on Catastrophic Nuclear Accidents.* Washington, D.C.: Presidential Commission on Catastrophic Nuclear Accidents, 1990. A report from a committee authorized by the Price-Anderson amendments of 1988 and appointed to investigate all aspects of possible accidents that might occur in U.S. nuclear power plants.

Stefenson, Bror, Per Axel Landahl, and Tom Ritchey. *Nuclear Accidents and Crisis Management:* Stockholm: Kungliga Krigsvetenskapskademien, 1993. A collection of papers prepared for a conference on this subject held in Sweden in 1993.

INTERNET/WEB DOCUMENTS

"Chernobyl Nuclear Disaster." Available online. URL: http://www.chernobyl. co.uk. Downloaded on February 9, 2005. A very complete web site that discusses the causes and consequences of the Chernobyl disaster, including a BBC report on the accident.

InfoUkes Inc. "The Chornobyl [sic] Nuclear Accident and It's [sic] Ramifications." Available online. URL: http://www.infoukes.com/history/ chornobyl. Downloaded on February 9, 2005. A collection of photographs, maps, reports, and commentaries on the Chernobyl nuclear disaster of 1986.

"List of Nuclear Accidents." Available online. URL: http://www.fact-index.com/l/li/list_of_nuclear_accidents.html. Downloaded on February 9, 2005. An exhaustive list of nuclear accidents throughout the world, organized by decades, with a list of unconfirmed accidents and some useful external links.

Lutins, Allen. "U.S. Nuclear Accidents." Available online. URL: http:// www.lutins.org/nukes.html. Downloaded on February 9, 2005. A very complete list of accidents that have occurred in U.S. nuclear power plants, with a brief commentary on each.

Meshkati, Najmedin. "Tokaimura (Japan's Sept. 30, 1999) Nuclear Accident: Its Causes, Health & Environmental Effects." Available online. URL: http://www-bcf.usc.edu/%7Emeshkati/tefall99/toki.html. Downloaded on February 9, 2005. Report of a project conducted by the students in a freshman class at the University of Southern California on a nuclear accident that occurred in Tokaimura, Japan, in 1999.

"Nuclear Reactor Accidents." Available online. URL: http://www.science. uwaterloo.ca/~cchieh/cact/nuctek/accident.html. Downloaded on February 9, 2005. An excellent introduction to the technical aspects of the accidents that occurred at Three Mile Island and Chernobyl along with a discussion of the nature of other kinds of nuclear accidents.

CURRENT ISSUES

BOOKS

Baker, Richard, ed. *Nuclear Terrorism.* Hauppauge, N.Y.: Nova Science Publishers, 2003. A series of articles discussing the vulnerability of nuclear power plants to terrorist attacks.

Beck, Peter. *Prospects and Strategies for Nuclear Power: Global Boon or Dangerous Diversion?* London: Royal Institute of International Affairs, Energy and Environmental Programme, 1994. The author examines the future of nuclear power, given the uncertainties of future energy technologies, the inadequacies of solutions for safety issues, and the reluctance of many individuals and organizations to embrace nuclear reactors as a solution to energy needs.

Benjamin-Alvarado, Jonathan. *Power to the People: Energy and the Cuban Nuclear Program.* London: Routledge, 2000. An analysis of the development of nuclear power in Cuba, a nation that would appear to be an unlikely candidate for such a development, with a consideration of the social, political, and economic consequences of Cuba's decision to move forward with nuclear power capability.

Bergeron, Kenneth D. *Tritium on Ice: The Dangerous New Alliance of Nuclear Weapons and Nuclear Power.* Cambridge, Mass.: MIT Press, 2003. A decision by the Clinton administration to start producing tritium for nuclear weapons in nuclear power plants raises some issues with potentially profound significance for both the military and peacetime applications of nuclear power.

Brooks, Lisa, and Jim Riccio. *Nuclear Lemons: An Assessment of America's Worst Commercial Nuclear Power Plants.* 5th edition. Washington, D.C.:

Public Citizen, 1996. A report prepared by Public Citizen's Critical Mass Energy Project that identifies the nation's 100-plus nuclear reactors on the basis of safety, economic efficiency, and performance. The report concludes that the Nuclear Regulatory Commission has been acting "with callous disregard for public health and safety."

Daley, Michael J. *Nuclear Power: Promise or Peril?* Minneapolis: Lerner Publishing, 1997. A "pro and con" book dealing with nuclear energy issues intended for young adult readers.

Dawson, Frank G. *Nuclear Power: Development and Management of a Technology.* Seattle: University of Washington Press, 1976. A revision of the author's Ph.D. thesis at the University of Washington that explores the issues involved in the development of an entirely new type of technology—the generation of electricity through nuclear reactions—and the mechanisms by which those issues were resolved.

Francis, John, and Paul Abrecht, eds. *Facing up to Nuclear Power: A Contribution to the Debate on the Risks and Potentialities of the Large-Scale Use of Nuclear Energy.* Edinburgh, Scotland: St. Andrew Press, 1976. A collection of essays by scientists, politicians, theologians, philosophers, and others on the impact that nuclear power is likely to make on human society in the future.

Freeman, Leslie J. *Nuclear Witnesses: Insiders Speak Out.* New York: W. W. Norton, 1981. Observations from a number of individuals who came to their jobs in the nuclear industry with great hopes for the promise of nuclear power but were later disillusioned by the way the industry operated.

Gerard, Michael B. *Whose Backyard, Whose Risk: Fear and Fairness in Toxic and Nuclear Waste Siting.* Cambridge, Mass.: MIT Press, 1996. The author, an environmental attorney, examines the economic, political, social, and psychological problems involved with selecting sites for nuclear waste storage and disposal and suggests certain principles by which such decisions can be made while providing some measure of fairness to all parties involved in such decisions.

Greenhalgh, Geoffrey. *The Future of Nuclear Power.* London: Graham and Trotman, 1998. Associated with the British nuclear association A Power for Good, the author attempts to assess the direction of nuclear power in the future, given its previous history.

Grimston, Malcolm C., and Peter Beck. *Civil Nuclear Energy: Fuel of the Future or Relic of the Past.* London: Royal Institute of International Affairs, 2001. An effort by the authors to eschew the emotional content of arguments about nuclear power plants and to set out objective reasons for both supporting and opposing such plants, with suggestions for ways by which a more rational debate can be conducted in the future.

Hill, C. R., A. L. Mechelynck, G. Ripka, and B. C. C. van der Zwaan. *Nuclear Energy: Promise or Peril?* Singapore: World Scientific Publishing,

1999. A comprehensive discussion of the potential benefits and risks posed by the widespread adoption of nuclear power as a source of energy in nations around the world, with special attention to its probable environmental benefits and the risks it poses of making more readily available the nuclear materials needed for the production of nuclear weapons.

Hodgson, P. E. *Nuclear Power, Energy and the Environment.* London: Imperial College Press, 1999. An analysis of the role of nuclear power in meeting the world's energy demands in the future, along with an analysis of the environmental effects of nuclear power and other forms of energy.

International Energy Agency. *Nuclear Power: Sustainability, Climate Change and Competition.* Paris: International Energy Agency, 1998. At a time when the growth of nuclear power worldwide has essentially come to a halt, this publication attempts to examine the issues that will determine whether nuclear energy continues its decline or once more begins to expand. The publication considers three areas in particular: environmental effects, economic viability, and competitive advantages and disadvantages of nuclear power. The publication is also available online at http://www.iea.org/dbtw-wpd/textbase/nppdf/free/1990/nuclearpower98.pdf.

Kaku, Michio, and Jennifer Trainer, eds. *Nuclear Power—Both Sides: The Best Arguments for and against the Most Controversial Technology.* New York: Norton, 1982. A somewhat dated book that still contains some of the most fundamental arguments for and against the use of nuclear reactors for the generation of power. Essays cover topics such as radiation, reactor safety, nuclear waste disposal, the economics of nuclear power generation, and the future of nuclear power.

Kursunoglu, Behram N., Stephan L. Mintz, and Arnold Perlmutter, eds. *The Challenges to Nuclear Power in the Twenty-First Century.* New York: Plenum Press, 2000. A collection of papers presented at the 22nd International Energy Forum, in which experts from a variety of fields discuss issues such as the need for nuclear energy in the future, competition between nuclear and other forms of energy, the interface of nuclear power and nuclear arms development, and the development of nuclear energy sources.

———. *Preparing the Ground for Renewal of Nuclear Power.* New York: Kluwer, 1999. A collection of papers presented at the International Conference on Preparing the Ground for Renewal of Nuclear Power, sponsored by the Global Foundation, held in Paris on October 22–23, 1998. The purpose of the conference was to consider a number of issues arising out of the fact that many first-generation nuclear power plants are now growing old and will need to be decommissioned or have their licenses renewed.

Leonard, Barry, ed. *Nuclear Power Plant Security: Voices from Inside the Fences.* Washington: Project on Government Oversight, 2003. A report on a POGO study on the safety of nuclear power plants, based to a large extent on interviews with workers at such plants responsible for safety procedures.

Nunn, Sam, and Robert E. Ebel. *Managing the Global Nuclear Materials Threat: Policy Recommendations.* Washington, D.C.: Center for Strategic & International Studies, 1999. Policy recommendations on five topics related to the use of nuclear materials, including methods for improving nuclear security, developing methods for dealing with nuclear wastes produced as the result of research and development in Russia, finding productive uses for excess nuclear materials produced by defense research and development, improving the safe management of nuclear materials worldwide, and maintaining the United States's current dominance in nuclear infrastructure.

Perin, Constance. *Shouldering Risks: The Culture of Control in the Nuclear Power Industry.* Princeton, N.J.: Princeton University Press, 2004. An investigation of the way in which potential problems are handled in nuclear power plants and why they are apparently so poorly resolved. The author interviews 60 experts in the area of nuclear safety to learn more about the reason the nuclear industry seems to have done such an inadequate job of providing the safe operation of its plants.

Porro, Jeffrey, et al., eds. *The Nuclear Age Reader.* New York: Knopf, 1989. A collection of articles, documents, and other materials dating from the early 1940s to the end of the 1980s, designed to be used in connection with the PBS series *War and Peace in the Nuclear Age.*

Risoluti, Piero. *Nuclear Waste: A Technological and Political Challenge.* New York: Springer-Verlag, 2004. An analysis of the technological and social problems involved in the disposal of nuclear wastes, the origins of public concerns about such wastes and waste disposal systems, and some methods that can be used to deal with those concerns.

Roberts, Alan, and Zhores Medvedev. *Hazards of Nuclear Power.* Nottingham, U.K.: Spokesman Books, 1977. An assessment of the environmental and human risks posed by the use of nuclear reactors for the generation of electrical power, published by the Bertrand Russell Peace Foundation.

Shapiro, Fred C. *Radwaste: A Reporter's Investigation of Nuclear Waste Disposal.* New York: Random House, 1981. A now-very-outdated story of the nation's nuclear waste disposal problems, still of value, however, because of the fundamental information it provides and the basic issues it defines and discusses.

Shrader-Frechett, K. S. *Burying Uncertainty: Risk and the Case against Geological Disposal of Nuclear Waste.* Berkeley: University of California Press,

1993. The author, professor of philosophy at the University of South Florida, argues that the risks of burying nuclear wastes underground are so great that such wastes should instead be stored above ground for at least 100 years to allow their constant monitoring.

Sims, Gordon H. E. *The Anti-Nuclear Game*. Ottawa, Canada: University of Ottawa Press, 1997. A critical analysis from a Canadian point of view of the objections that have been raised to the construction of nuclear power plants, with arguments supporting nuclear power as an important source of energy in the world's future.

Thomas, Steve D., Chris Hope, and Jim Skea. *The Realities of Nuclear Power: International Economic and Regulatory Experience*. Cambridge: Cambridge University Press, 1988. Declining orders for nuclear power plants in many parts of the world provide a motivation for reassessing the potential of nuclear energy as a source of power throughout the world and an analysis of the reason that nuclear power plants have been utilized successfully in some countries of the world but not in others.

Utgoff, Victor, A, ed. *The Coming Crisis: Nuclear Proliferation, U.S. Interests, and World Order*. Cambridge, Mass.: MIT Press, 2000. An analysis of the political and social consequences of the proliferation of nuclear weapons development to a variety of nations throughout the world and what this development can mean for the stability of world order in the next generation.

Weber, Isabelle P., and Susan D. Wiltshire. *The Nuclear Waste Primer: A Handbook for Citizens*. New York: Schocken Books, 1985. An unbiased presentation of the facts relating to nuclear wastes, along with suggestions as to how citizens can become involved in the effort to find solutions to this problem.

Weiss, Ann E. *The Nuclear Question*. New York: Harcourt, 1981. A general introduction to the nature of nuclear energy and the controversial issues surrounding the use of nuclear reactors.

Williams, David R. *What Is Safe?: The Risks of Living in a Nuclear Age*. Cambridge, U.K.: Royal Society of Chemistry, 1998. The author discusses some fundamental issues concerning the nature of risk, various sources and types of risk, and risk management, with special attention to the risks that result from the use of nuclear energy in our society.

Wilpert, Bernhard, and Naosuke Itoigawa, eds. *Safety Culture in Nuclear Power Operations*. London: Taylor & Francis, 2001. Experts in a variety of fields discuss the scientific, technical, social, economic, and psychological aspects of nuclear power plant safety.

Wolfson, Richard. *Nuclear Choices: A Citizen's Guide to Nuclear Technology*. Revised edition. Cambridge, Mass.: MIT Press, 1993. An effort to provide readers with the technical background about nuclear energy that will allow them to make decisions as to its use in both military and peaceful applications, with particular attention to the benefits that derive from its value as a source of electrical power in the modern world.

MAGAZINES AND JOURNALS

Alvarez, Robert. "What about the Spent Fuel?" *Bulletin of the Atomic Scientists*, vol. 58, no. 1, January/February 2002, pp. 45–47. The author points out that nuclear power plants are vulnerable to terrorist attacks not simply because of the plants themselves, but also because of the spent fuel wastes that are generally stored onsite near the plants.

Bergeron, Kenneth. "While No One Was Looking." *Bulletin of the Atomic Scientists*, vol. 57, no. 2, March/April 2001, pp. 42–49. It has long been U.S. policy not to use commercial nuclear power plants for the purpose of producing weapons-grade nuclear materials, but that policy has apparently been violated in practice on more than one occasion.

"Bouillabaisse Sushi." *The Economist*, vol. 370, no. 8361, February 5, 2004, pp. 74–75. An observation on international plans to build a giant new experimental fusion reactor and why those plans may not be such a good idea.

Burke, Tom. "Uranium Dangers." *New Statesman*, vol. 16, no. 745, February 24, 2003, pp. R6–7. A discussion of the military risks in the development of nuclear power plants in that the fuels and other materials used for such plants can generally be adapted for weapons use.

Butler, Declan. "Nuclear Power's New Dawn." *Nature*, vol. 429, May 20, 2004, pp. 238–240. Concerns about climate change and increases in energy demand appear to make nuclear power a more attractive alternative, but safety and cost issues continue to plague the industry.

Chapin, Douglas M., et al. "Nuclear Power Plants and Their Fuel as Terrorist Targets." *Science*, vol. 297, issue 5589, September 20, 2002, pp. 1,997–1,999. Following the September 11, 2001, attack on the World Trade Center, a number of questions were raised about the safety of nuclear power plants from terrorist attacks. This article provides a statement prepared by a committee of the National Academy of Engineers outlining an objective analysis of that risk.

Cohn, Laura, John Carey, and Michael Arndt. "A Comeback for Nukes?" *Business Week*, Issue 3729, April 23, 2001, p. 38. The upswing in interest in nuclear power plants can be attributed in part to greater efficiency and economic viability of the technology along with the availability of new technology, such as the pebble bed reactor, but the safety of nuclear power plants is still a serious concern for consumers.

Cravens, Gwyneth. "Terrorism and Nuclear Energy: Understanding the Risks." *Brookings Review*, vol. 20, Spring 2002, pp. 40–44. The terrorist attacks of September 11, 2001, have raised the concerns of Americans about possible nuclear attacks in the future. In order to assess these risks, we need a better understanding of the nature of nuclear energy and the threats it may pose.

Crowley, Kevin D. "Nuclear Waste Disposal: The Technical Challenges." *Physics Today*, vol. 50, June 1997, pp. 32–39. A valuable article because it lays out in clear terms most of the fundamental issues involved in nuclear waste disposal, issues that are essentially the same today as they were in 1997. The article was part of a special issue on nuclear wastes.

Domenici, Pete V. "Future Perspectives on Nuclear Issues." *Issues in Science and Technology*, vol. 14, Winter 1997, pp. 53–59. The author argues that the nation needs to take a fresh look at nuclear technologies, especially in view of our profound dependence on foreign nations for energy reserves, such as oil. He also states that policymakers have done a relatively poor job of assessing risks and benefits of nuclear power thus far and need to improve their ability to assess these factors.

Eisenberg, Daniel. "Nuclear Summer." *Time*, vol. 157, no. 21, May 28, 2001, pp. 58–60. The Bush administration is promoting the expansion of nuclear power in the United States, but its success in this effort will be determined largely by economic considerations.

Ewing, Rodney C., and Allison Macfarlane. "Yucca Mountain." *Science*, vol. 296, issue 5568, April 26, 2002, pp. 659–660. The authors argue that political pressures have become so great over nuclear waste disposal issues that it is those pressures, rather than scientific information, that is driving decisions as to where and how radioactive wastes are to be disposed.

Farber, Darryl, and Jennifer Weeks. "A Graceful Exit? Decommissioning Nuclear Power Reactors." *Environment*, vol. 43, no. 6, July 2001, pp. 8–21. A discussion of the environmental risks involved in the decommissioning of old nuclear reactors and policy decisions that can be made to reduce or deal with those risks.

Freemantle, Michael. "Nuclear Power for the Future." *Chemical & Engineering News*, vol. 82, no. 37, September 13, 2004, pp. 31–35. A discussion of the way in which new reactor designs appear to give promise for operating nuclear power plants with greater safety, at greater efficiency, and with simpler technological designs.

Hertsgaard, Mark. "Three Mile Island." *The Nation*, vol. 278, no. 13, March 18, 2004, p. 7. During the terrorist attacks on the Twin Towers in New York City on September 11, 2001, guards at the Three Mile Island nuclear power plant attempted to activate the plant's security system designed to protect against such attacks, only to discover that the system didn't work. The author considers what this event means in the greater context of President Bush's commitment to increasing the nation's dependence on nuclear power plants.

Hirsch, Daniel. "The NRC: What, Me Worry?" *Bulletin of the Atomic Scientists*, vol. 58, no. 1, January/February 2002, pp. 38–44. Following the September 11, 2001, attack on the World Trade Center, the Nuclear Regulatory Commission announced that the United States's nuclear power plants were safe against such attacks, but the author argues that the NRC's safety requirements are low and its confidence in plant safety is probably misguided.

Hirsch, Daniel, David Lochbaum, and Edwin Lyman. "The NRC's Dirty Little Secret." *Bulletin of the Atomic Scientists*, vol. 59, no. 3, May/June 2003, pp. 44–51. A discussion of the threat posed by a terrorist attack at a nuclear power plant and the Nuclear Regulatory Agency's apparent unwillingness to take such threats seriously.

Johnson, Jeff. "New Life for Nuclear Power?" *Chemical & Engineering News,"* October 2, 2000, pp. 39–43. An analysis of the status of nuclear power generation in the United States with a discussion of the way in which changes made by the NRC in licensing regulations has promoted a renewed interest in nuclear power.

———. "Who's Watching the Reactors?" *Chemical & Engineering News*, vol. 81, no. 19, May 12, 2003, pp. 27–31. A near-major accident at the Davis Besse Nuclear Power Station in Oak Harbor, Ohio, raises questions as to whether the NRC is monitoring the operation and safety of nuclear power plants adequately.

Keenan, Charlie. "Nuclear Nightmare?" *Science World*, vol. 58, no. 11, March 11, 2002, pp. 17–19. The author explores how great a risk nuclear power plants pose for possible terrorist attacks in the United States. The article is written for high school students.

Lanouette, William. "Greenhouse Scare Reheats Nuclear Debate." *Bulletin of the Atomic Scientists*, vol. 46, no. 3, April 1990, p. 34. Concerns about global climate change have encouraged political leaders to think more seriously about the use of nuclear power as a source of energy to replace fossil fuel combustion, but such hopes may be overly optimistic.

Laurent, Christine. "Beating Global Warming with Nuclear Power?" *UNESCO Courier*, vol. 54, no. 2, February 2001, pp. 37–40. The author outlines arguments for the use of nuclear power plants to reduce greenhouse gas emissions and suggests some counterarguments to these viewpoints. A pair of pro and con statements follow the article.

Leon, J. L.-L. "Radioactive Waste Management and Sustainable Development." *NEA News*, vol. 19, no. 1, January 2000, pp. 18–20. A discussion of the problems of energy generation in developing nations and the criteria that need to be applied to select among various sources of energy. Nuclear power may be one of the most promising of those sources.

Annotated Bibliography

Leon, J. L.-L., C. Picot, and H. Riotte. "Sustainable Solutions for Radioactive Waste." *OECD Observer*, no. 226–227, Summer 2001, pp. 18–19. Nuclear power is a promising technology in the battle against greenhouse gas emissions, but the problem of waste disposal must be solved before the technology can be widely implemented.

Leventhal, Paul L., and Milton M. Hoenig. "Nuclear Terrorism—Reactor Sabotage and Weapons Proliferation Risks." *Contemporary Policy Issues*, vol. 8, no. 3, Fall 1990, pp. 106–121. Some issues involved in possible terrorist attacks on nuclear power plants and a discussion of problems associated with the proliferation of nuclear weapons.

Levi, Michael A. "The Wrong Way to Promote Nuclear Power." *The New Republic*, November 24, 2003, pp. 7–8. The author points out that, at a time when nuclear power plants are being phased out in Europe, they seem to be experiencing a renaissance in the United States. He examines the reasons for these opposing trends and explains why it is that both parts of the world may not be "getting it right."

Long, Michael. "Half Life: The Lethal Legacy of America's Nuclear Waste." *National Geographic*, vol. 202, no. 1, January 2002, pp. 1–33. The author spent six weeks visiting nuclear power plants throughout the nation as background for this general introduction to the problems involved with the generation and disposal of nuclear wastes.

Malakoff, David. "Nuclear Power: New DOE Research Program to Boost Sagging Industry." *Science*, vol. 282, issue 5396, December 11, 1998, pp. 1,980–1,981. An announcement of the Department of Energy's new program to promote the development of nuclear reactors, the Nuclear Energy Research Initiative (NERI).

Nadis, Steven. "The Sub-Seabed Solution." *The Atlantic Monthly*, vol. 278, issue 4, October 1996, pp. 28–31. Geologist Charles Hollister offers a proposal for burying nuclear wastes beneath the ocean's bottoms, but the plan may never receive a legitimate evaluation for political reasons.

O'Meara, Kelly Patricia. "Nuclear Power Rises Again." *Insight on the News*, vol. 17, June 11, 2001, pp. 12–13. The new Republican administration is surprising energy experts by suggesting efforts to decrease dependence on traditional fossil fuels by re-energizing the nation's nuclear program. President Bush suggests that 300 new nuclear power plants should be constructed.

"Report on Nuclear Terrorism." *Bulletin of the Atomic Scientists*, vol. 42, December 1986, pp. 38–44. Excerpts from a report on nuclear terrorism conducted by the International Task Force on Prevention of Nuclear Terrorism.

REPORTS

Beck, Peter, and Malcolm Grimston. *Double or Quits? The Global Future of Civil Nuclear Energy*. London: The Royal Institute of International Affairs,

April 2002. An analysis of the benefits and risks presented by nuclear power and the likelihood that it will be able to meet some portion of the world's energy needs in the future. The report is also available online at http://www.acus.org/Energy/Nuclear_Double_or_Quits.pdf.

Committee on Future Nuclear Power Development. *Nuclear Power: Technical and Institutional Options for the Future.* Washington, D.C.: National Academies Press, 1992. The report of a committee of the Energy Engineering Board of the National Research Council appointed to consider the current status of reactor technology in the United States, future prospects for nuclear power plants, and possible alternatives to existing reactor designs.

The Future of Nuclear Power: An Interdisciplinary MIT Study. Cambridge, Mass.: Massachusetts Institute of Technology, 2003. A comprehensive study by a committee of faculty members at MIT who believe that nuclear power may be an important factor in the United States's future energy equation, provided that four key issues can be resolved: cost, safety, waste, and proliferation. The report is also available online at http://web.mit.edu/nuclearpower.

Kotek, J. F., et al. *Nuclear Energy: Power for the Twenty-First Century.* Oak Ridge, Tenn.: U.S. Department of Energy, Office of Scientific and Technical Information, May 2003. A report prepared by scientists at Oak Ridge, Los Alamos, Sandia, Argonne, and Lawrence Livermore National Laboratories, and the Idaho National Engineering and Environmental Laboratory on the future of nuclear power in the United States. The report is also available online at http://nuclear.inel.gov/papers-presentations/power_for_the_21st_century.pdf.

INTERNET/WEB DOCUMENTS

British Broadcasting Corporation. "Chernobyl Nuclear Disaster." Available online. URL: http://www.chernobyl.co.uk. Downloaded on February 10, 2005. Transcript of a broadcast by the BBC on the Chernobyl nuclear power plant accident of 1986.

European Commission. "Nuclear Issues," Available online. URL: http://europa.eu.int/comm/energy/nuclear/index_en.html. Downloaded on February 10, 2005. An excellent web site that discusses a variety of nuclear issues, such as radioactive wastes, decommissioning of nuclear plants, radiation safety issues, and the transport of nuclear materials, all from a European standpoint.

Frontline. "Nuclear Reaction: Why Do Americans Fear Nuclear Power?" PBS online. URL: http://www.pbs.org/wgbh/pages/frontline/shows/reaction. Downloaded on February 10, 2005. Provides readings, maps, charts, interviews, nuclear links, press reactions, nuclear glossary, discussions of safety issues, and other information on nuclear power plants.

Handwerk, Brian. "Nuclear Terrorism—How Great Is the Threat?" National Geographic News, Available online. URL: http://news.nationalgeographic. com/news/2002/10/1011_021011_nuclear.html. Downloaded on February 10, 2005. A discussion of the variety of ways in which nuclear terrorism might be carried out and an assessment of the relative risk posed by such possibilities.

Holt, Mark, and Carl E. Behrens. "Nuclear Energy Policy." Available online. URL: http://www.cnie.org/nle/crsreports/energy/eng-5.pdf. Downloaded on February 12, 2005. A report by the Congressional Research Service, published on March 22, 2001, providing an extended policy analysis of the use of nuclear energy for power production and weapons manufacture in the United States and other nations of the world.

Massachusetts Institute of Technology. "The Future of Nuclear Power." Available online. URL: http://web.mit.edu/nuclearpower. Downloaded on February 12, 2005. A committee of scholars at the Massachusetts Institute of Technology reports on "the most comprehensive, interdisciplinary study ever conducted on the future of nuclear energy."

"Nuclear Energy News." Available online URL: http://www.topix.net/tech/ nuclear-energy. Downloaded on February 12, 2005. Current articles about nuclear energy from newspapers around the world on a web site that calls itself "The Internet's Largest News Site."

"Nuclear Energy Policy." Available online. URL: http://fpc.state.gov/ documents/organization/36307.pdf. Downloaded on February 12, 2005. A very informative summary of the nation's nuclear energy policy on power production, liability, waste disposal, and other issues prepared in September 2004 for the Congress by the Congressional Research Service of the Library of Congress.

Project on Government Oversight. "Nuclear Power Plant Security: Voices from Inside the Fences." Available online. URL: http://www.pogo.org/p/ environment/eo-020901-nukepower.html. Downloaded on February 12, 2005. Special report by a private, nonprofit organization whose objective it is to investigate, expose, and suggest remedies for "abuses of power, mismanagement, and subservience by the federal government to powerful special interests."

Settle, Frank. "Nuclear Chemistry and the Community." Available online. URL: http://www.chemcases.com/nuclear. Downloaded on February 12, 2005. One of 12 case studies developed at Kennesaw State University for second-semester college chemistry students with support from the National Science Foundation. The two 45-minute lessons cover historical topics, chemistry of the uranium fuel cycle, nuclear reactors, and waste disposal issues.

Three Mile Island Alert, Inc. "Nuclear Terrorism: Sabotage and Terrorism of Nuclear Power Plants." Available online. URL: http://www.tmia.com/sabter.html. Downloaded on February 12, 2005. A very complete web site that discusses many aspects of the potential threat of terrorism on nuclear power plants, with a number of useful links to related web pages and other resources.

SUPPORT FOR NUCLEAR POWER

BOOKS

Beckman, Petr. *The Health Hazards of Not Going Nuclear.* Boulder, Colo.: Golem Press, 1977. The author argues that most forms of energy production present a greater hazard to human health than does nuclear power generation, and that, therefore, greater emphasis should be placed on the development of that energy source.

Domenici, Pete V. *A Brighter Tomorrow: Fulfilling the Promise of Nuclear Energy.* Lanham, Md.: Rowan & Littlefield, 2004. The senior senator from New Mexico argues that nuclear power will play an increasingly larger role in the American energy equation in the future and why that is a positive step for the nation.

Greenhalgh, Geoffrey, ed. *The Necessity for Nuclear Power.* London: Graham and Trotman, 1990. A collection of essays outlining the arguments for an increased use of nuclear power plants in today's world.

Heaberlin, Scott W. *Case for Nuclear-Generated Electricity, or, Why I Think Nuclear Power Is Cool and Why It Is Important that You Think So Too.* Columbus, Ohio: Battelle Press, 2003. A strong argument for the development of nuclear power as a major energy source for the future in the United States and the rest of the world.

McCracken, Samuel. *The War against the Atom.* New York: Basic Books, 1982. A strong proponent of nuclear power plants subtitles his book "The Overwhelming Case for Nuclear Power—and against the Groups and Individuals who Continue to Fight It."

Morris, Robert C. *The Environmental Case for Nuclear Power: Economic, Medical, and Political Considerations.* New York: Paragon House, 2000. A retired chemistry teacher argues that opponents of nuclear power plants overstate the risk they pose to humans and the environment and that their contributions to society are actually quite substantial.

Waltar, Alan E. *America the Powerless: Facing Our Nuclear Energy Dilemma.* Madison, Wisc.: Medical Physics Publishing, 1996. A former present of the American Nuclear Society argues that America's unwillingness to develop nuclear power more aggressively places us at the mercy of other nations around the world, from whom we have to import oil, coal, and other energy resources.

MAGAZINES AND JOURNALS

Gerholm, Tor Ragnar. "The Atomic Age Is Not Over Yet," *New Statesman*, vol. 11, no. 523, September 25, 1998, p. S22. Environmentalists should begin to accept the fact that nuclear power is here to stay and that no other environmentally friendly method of power production is likely to become widely available in the foreseeable future.

Greenwald, John. "Time to Choose." *Time*, vol. 137, April 29, 1991, pp. 54–61. The author admits that the development of nuclear power is problematic because of some serious risks posed by the technology, but he argues that no other method of producing energy is likely to be able to solve America's energy needs in the future.

Jaworowski, Zbigniew. "Radiation Risk and Ethics." *Physics Today*, vol. 52, no. 9, September 1999, pp. 24–29. The author argues that the risk from low-level doses of radiation is exaggerated and the public should be told they are not at risk from the small amount of radiation received from nuclear power plants, fallout at great distances, and other sources.

Pryor, Charles. "The Vital Role of Nuclear Energy." *New Statesman*, vol. 16, no. 745, February 24, 2003, p. R17. The author argues for nuclear power as an essential part of the world's "energy portfolio"; that it is now a safe, reliable, and economically competitive source of energy; and that the disposal of wastes remains the most serious problem related to nuclear power production.

Wolfe, Bertram. "Why Environmentalists Should Promote Nuclear Energy." *Issues in Science and Technology*, vol. 12, Summer 1996, pp. 55–60. Increasing demands for energy require that policymakers examine all possible alternative sources. Given nuclear energy's excellent safety record, especially in comparison with other energy sources, greater attention has to be paid to nuclear power plants with their promise of reducing the environmental harm produced by energy generation.

INTERNET/WEB DOCUMENTS

Adams, Rod. "Atomic Insights." Available online. URL: http://www. atomicinsights.com/AEI_home.html. Downloaded on February 10, 2005. A collection of newsletters begun in April 1995 as an "alternative source of information" about nuclear power.

"Environmentalists for Nuclear Energy." Available online. URL: http:// www.ecolo.org/base/baseus.htm. Downloaded on February 10, 2005. An international nonprofit organization of individuals interested in environmental issues who see nuclear power as a constructive factor in protecting the Earth's environment. The web site contains documents, news, photographs, letters, and other information relating to nuclear power use.

McCarthy, John. "Frequently Asked Questions about Nuclear Energy." Available online. URL: http://www.formal.stanford.edu/jmc/progress/nuclear-faq.html. Downloaded on February 10, 2005. The author of this page, Professor Emeritus of Computer Science at Stanford University, touts nuclear energy as an important factor in the development of a sustainable economy in both the United States and the rest of the world.

Smith, Roy. "Nuclear Power: Energy for Today and Tomorrow." Available online. URL: http://pw1.netcom.com/~~res95/energy/nuclear.html. Downloaded on February 10, 2005. A web site intended to "educate citizens about the environmental benefits of nuclear power" with a number of links to governmental and industrial sites with further information about nuclear power.

OPPOSITION TO NUCLEAR POWER

BOOKS

Caldicott, Helen. *Nuclear Madness: What You Can Do.* Revised edition. New York: W. W. Norton, 1994. The founder of Physicians for Social Responsibility and Women's Action for Nuclear Disarmament has revised and republished her classic 1978 book warning about the dangers of nuclear power production.

Dickerson, Carrie Barefoot, and Patricia Lemon. *Aunt Carrie's War against Black Fox Nuclear Power Plant.* Portland, Ore.: Council Oak Distribution, 1995. A recounting of the efforts by Dickerson, a registered nurse and owner of Aunt Carrie's Nursing Home in Claremore, Oklahoma, to prevent the construction of a nuclear power plant at Inola, near her home. Dickerson was successful in her efforts, and Oklahoma remains one of the few states without a nuclear reactor for the commercial generation of power.

Environmental Action Foundation. *Accidents Will Happen: The Case against Nuclear Power.* New York: Harper & Row, 1979. A strong indictment of the nuclear power industry and a call for ending the construction of new nuclear power plants.

Falk, Jim. *Global Fission: The Battle over Nuclear Power.* Melbourne: Oxford University Press, 1983. A strong opponent of nuclear power analyzes the growth of resistance to nuclear reactors throughout the world, with special attention to the antinuclear movement in his home country of Australia.

Flam, Helena, ed. *States and Anti-Nuclear Movements.* Edinburgh, Scotland: Edinburgh University Press, 1994. Authors from eight Western European nations analyze the formation, evolution, policies, and actions of organizations that have developed to oppose the construction of nuclear power plants in their own countries.

Foreman, Harry, ed. *Nuclear Power and the Public.* Minneapolis: University of Minnesota Press, 1970. A review of the birth, growth, and activities of groups opposed to the development of nuclear power plants.

Freidel, Frank, and Ernest Hay, eds. *Ideology, Interest Group Formation and the New Left: The Case of the Clamshell Alliance.* New York: Garland Publishing, 1988. Based on the author's Ph.D. thesis at Harvard University, this book outlines the philosophical basis underlying arguably the most effective of all single-site anti-nuclear groups formed to fight nuclear plant construction and a history of its work during the 1970s.

Gofman, John W., and Arthur R. Tamplin. *Poisoned Power: The Case against Nuclear Power Plants before and after Three Mile Island.* Emmaus, Pa.: Rodale Press, 1979. Gofman, sometimes called the Father of the Antinuclear Movement in the United States, presents a strong case against the development of nuclear power because of the risks it poses to human health and survival and to the environment.

Goodman, Sidney. *Asleep at the Geiger Counter: Nuclear Destruction of the Planet and How to Stop It.* Nevada City, Calif.: Blue Dolphin Publishing, 2002. A work that claims to document "why nuclear programs mean cancer everywhere, birth defects forever, the uncontrolled proliferation of nuclear weapons, increased chance of nuclear war, the undermining of our national defense, the loss of civil liberties, and unending multi-billion dollar subsidies which have been bleeding us white, while actually delaying genuine energy independence."

Grossman, Karl. *Cover Up: What You Are Not Supposed to Know about Nuclear Power.* Sagaponack, N.Y.: Permanent Press, 1980. Grossman argues that nuclear power plants are not safe and should not be constructed.

———. *The Wrong Stuff: The Space Program's Nuclear Threat to Our Planet.* Monroe, Me.: Common Courage Press, 1997. A critique of the U.S. government's efforts to expand its use of nuclear materials in space as, for example, a source of power in the 1997 Cassini space mission to Saturn.

Gyorgy, Anna, et al. *No Nukes: Everyone's Guide to Nuclear Power.* Boston: South End Press, 1979. Widely regarded as one of the "classics" in the antinuclear movement, released shortly after the Three Mile Island accident.

Hertsgaard, Mark. *Nuclear Inc.: The Men and Money behind Nuclear Energy.* New York: Pantheon Books, 1983. A classic, muckraking exposé of the corporations and individuals who operate the nuclear power industry, their vision for the role of nuclear power in the United States and the world, and the consequences that may arise as a result of this vision.

Hilgartner, Stephen, Richard C. Bell, and Rory O'Conner. *Nukespeak: Nuclear Language, Visions, and Mindset.* San Francisco: Sierra Club Books, 1982. A strong condemnation of the nuclear power movement with a useful introduction to the political and economic aspects of nuclear power plant development.

Joppke, Christian. *Mobilizing against Nuclear Energy: A Comparison of Germany and the United States.* Berkeley: University of California Press, 1993. An analysis of comparison of the political movements against the development of nuclear power plants in two nations.

Lewis, Richard S. *The Nuclear-Power Rebellion: Citizens vs. the Atomic Industrial Establishment.* New York: Viking Press, 1972. A review and analysis of the development of an antinuclear power movement in the United States.

Makhijani, Arjun, and Scott Saleska. *The Nuclear Power Deception: U.S. Nuclear Mythology from Electricity "Too Cheap to Meter" to "Inherently Safe" Reactors.* New York: Apex Press, 1999. An analysis of the history of "wildly optimistic public statements" that have been made about nuclear power from the 1940s to the present day with some suggestions for ways of dealing with civilian and military nuclear issues today.

Nader, Ralph, and John Abbotts. *The Menace of Atomic Energy.* New York: Norton, 1977. The authors start by explaining the operation of a nuclear power plant and then point out with considerable force their reasons for opposing the construction and operation of such facilities.

Price, Jerome. *The Antinuclear Movement.* Boston: Twayne, 1982. The author analyzes the social and psychological factors involved in the development of groups organized in opposition to the development of nuclear power plants in the United States.

Pringle, Peter, and James Spigelman. *The Nuclear Barons.* London: Michael Joseph, 1981. Described as "the chilling story of how a small band of scientists, generals, politicians, and businessmen created the life-and-death issue confronting us today . . . our nuclear world."

Purcell, Arthur H., and Albert Fritsch. *Critical Hour: The End of the Road for Nuclear Energy?* Washington, D.C.: National Press Books, 1993. The authors suggest that ongoing problems related to the nuclear power industry may have doomed the future of reactor development in the United States.

Rudig, Wolfgang. *Anti-Nuclear Movements: A World Survey of Opposition to Nuclear Energy.* Harlow, U.K.: Longman Current Affairs, 1990. An analysis of similarities and differences among antinuclear movements in a number of countries throughout the world, with a review of the successes and failures of those movements.

Annotated Bibliography

Smith, Jennifer, ed. *The Antinuclear Movement*. Farmington Hills, Mich.: Greenhaven Press, 2002. A series of articles that discuss the history of the antinuclear movement, beginning with opposition to the development of the atomic and hydrogen bombs, and continuing through present-day campaigns aimed at both military and peaceful uses of nuclear energy.

Wellock, Thomas Raymond. *Critical Masses: Opposition to Nuclear Power in California, 1958–1978*. New York: Critical Press, 1998. An extraordinary historical analysis of the movement that developed in California and that was opposed to nuclear power plants and how that movement achieved such remarkable success.

MAGAZINES AND JOURNALS

Downey, Gary L. "Ideology and Clamshell Identity: Organizational Dilemmas in the Anti-nuclear Movement." *Social Problems*. vol. 33, no. 5, June 1986, pp. 357–373. An in-depth study of one of the most famous of the antinuclear groups, the Clamshell Alliance, with an analysis of the social and philosophical principles that motivated the movement.

Futrell, Robert, and Barbara G. Brents. "Protest as Terrorism? The Potential for Violent Anti-Nuclear Activism." *American Behavioral Scientist*, vol. 46, no. 6, June 2003, pp. 745–765. The authors explore the possibility of violent action against the federal government's plans to build a waste disposal site at Yucca Mountain, Nevada. They review the past history of such actions by antinuclear activists.

Kitschelt, Herbert P. "Political Opportunity Structures and Political Protest: Anti-nuclear Movement in Four Democracies," *British Journal of Political Science*, vol. 16, no. 1, Spring 1986, pp. 58–95. An extensive study of the birth and development of antinuclear movements in four nations—France, Great Britain, the United States, and West Germany—with an analysis of their impact on the political structure and decisions in each nation.

Mitchell, Robert Cameron. "From Elite Quarrel to Mass Movement." *Society*, vol. 18, no. 5, May 1981, pp. 76–84. A scholarly analysis of the early years of the antinuclear movement and how it changed from a relatively small and exclusive movement into one that included a wide range of individuals with many different concerns about nuclear power.

PAMPHLETS AND BROCHURES

Halstead, Fred. *What Working People Should Know about the Dangers of Nuclear Power*. Atlanta, Ga.: Pathfinder Books, 2nd edition, 2001. An explanation as to why working people should insist that all nuclear power plants be shut down immediately and why the problem of nuclear waste disposal is one that will threaten humans for thousands of years into the future.

INTERNET/WEB DOCUMENTS

"Bulletin of the Atomic Scientists." Available online. URL: http://www. thebulletin.org. Downloaded on February 10, 2005. Web page of the oldest and probably most highly respected critics of the use of nuclear power for both military and many peacetime applications.

The Canadian Coalition for Nuclear Responsibility. Available online. URL: http://www.ccnr.org. Downloaded on February 10, 2005. An extensive review of the physics, chemistry, and biology of nuclear materials by an organization opposed to the development of nuclear power plants.

Cohen, Bernard L. "Risks of Nuclear Power." Available online. URL: http://www.physics.isu.edu/radinf/np-risk.htm. Downloaded on February 10, 2005. An overview of the risks to human health posed by nuclear radiation from a variety of sources by one of the world's leading authorities on the subject.

"Energy Probe." Available online. URL: http://www.energyprobe.org. Downloaded on February 10, 2005. Homepage of a "consumer and environmental research team, active in the fight against nuclear power, and dedicated to resource conservation, economic efficiency, and effective utility regulation."

Makhijani, Arjun, and Scott Saleska. "The Nuclear Power Deception: U.S. Nuclear Mythology from Electricity 'Too Cheap to Meter' to 'Inherently Safe' Reactors." Available online. URL: http://www.ieer.org/reports/npd.html. Downloaded on February 10, 2005. A report issued in April 1996 outlining technical, social, and economic arguments against the expansion of nuclear power production.

"Mothers' Alert." Available online. URL: http://www.mothersalert.org. Downloaded on February 10, 2005. A web page devoted to providing information about the link between radiation and human health and actions that can be taken to reduce and eliminate nuclear threats.

National Environmental Coalition of Native Americans. Available online. URL: http://oraibi.alphacdc.com/necona. Downloaded on February 10, 2005. Web page for a Native American group interested in educating tribal members about the dangers of nuclear wastes and working to prevent the use of Native lands for dumping of such wastes.

"Nuclear Power Campaign." Available online. URL: http://www.greenpeace. org.uk/contentlookup.cfm?&SitekeyParam=D-E. Downloaded on February 10, 2005. An overview of the campaign against nuclear power plant by Greenpeace, one of the world's oldest and largest environmental organizations.

Thorpe, Grace. "Our Homes Are Not Dumps: Creating Nuclear-Free Zones." Available online. URL: http://www.alphacdc.com/necona/homes. Downloaded on February 10, 2005. An article by the daughter of Olympic champion Jim Thorpe arguing against the use of Native American lands for the dumping of nuclear wastes.

"U.S. Nuclear Plants in the 21st Century: The Risk of a Lifetime." Available online. URL: http://www.ucsusa.org/clean_energy/nuclear_safety/page.cfm?pageID=1408. Downloaded on February 10, 2005. Summary of a report by the Union of Concerned Scientists about the risks posed by nuclear power plants to the environment and human health.

LEGAL ISSUES

BOOKS

Ahearne, John F. *The Regulatory Process for Nuclear Power Reactors: A Review.* Washington, D.C.: Center for Strategic & International Studies, 1999. The report of a study conducted by the CSIS on the regulatory policies and practices under which the Nuclear Regulatory Commission monitors nuclear power plants in the United States. The study involved representatives of the nuclear industry as well as a variety of public interest groups.

Byrne, John, and Steven M. Hoffman. *Governing the Atom: The Politics of Risk.* Somerset, N.J.: Transaction Publishers, 1996. The development of nuclear energy over the past half century has resulted in and been accompanied by the rise of a variety of new social agencies designed for the control and administration of nuclear energy. This book examines the nature of such agencies in France, Germany, Russia, Eastern Europe, Korea, and Japan.

Cameron, Peter, Leigh Hancher, and Wolfgang Kuhn, eds. *Nuclear Energy Law after Chernobyl.* London: Graham & Trotman, 1988. A collection of papers that discuss the legal consequences that developed in Europe and elsewhere as a result of the Chernobyl disaster. One contributor observes that little changed as a result of the accident and, should a similar event occur at the time, the results would be largely the same as those observed two years earlier.

Critical Mass Energy Project. *Hear No Evil, See No Evil, Speak No Evil: What the NRC Won't Tell You about Nuclear Reactors.* Washington, D.C.: Public Citizen, 1993. A report by the Critical Mass Energy Project of Public Citizen that claims, based on unreleased documents from the Nuclear

Regulatory Commission, that nuclear power plants are far less safe than the general public is led to believe.

Curtis, Richard. *Perils of the Peaceful Atom: The Myth of Safe Nuclear Power Plants*. New York: Arco Publishing, 2000. The author suggests a number of reasons that nuclear power plants are not nearly as safe as the industry would have the general public believe and outlines some of the risks they pose to humans and the environment.

Curtis, Richard, and Elizabeth Hogan. *Nuclear Lessons: An Examination of Nuclear Power's Safety, Economic and Political Record*. Harrisburg, Pa.: Stackpole Books, 1980. The authors look with dissatisfaction on the safety record of the nuclear power industry.

Johnson, John W. *Insuring against Disaster: The Nuclear Industry on Trial*. Macon, Ga.: Mercer University Press, 1986. A legal analysis of the disputes over nuclear power, with special attention to the 1978 case between Duke Power Company and the Nuclear Regulatory Commission on one side and the Carolina Environmental Study Group on the other (1978). The author also discusses the Price-Anderson Act that limits industry's liability in case of a nuclear accident.

Mazuzan, George, and J. Samuel Walker. *Controlling the Atom: The Beginnings of Nuclear Regulation, 1946–1962*. Berkeley: University of California Press, 1984. The first of two books (see also J. Samuel Walker) providing a historical overview of the development of the Atomic Energy Commission and its successor agencies by a former historian of the National Science Foundation (Mazuzan) and a former historian of the Nuclear Regulatory Commission (Walker).

McCallion, Kenneth F. *Shoreham and the Rise and Fall of the Nuclear Power Industry*. New York: Praeger Publishers, 1995. A recounting of the problems involved in the effort by the Long Island Lighting Company to construct a nuclear power plant at Shoreham, New York, that ultimately resulted in a court case and a decision of mismanagement and fraud against the company that led to a multimillion dollar verdict and cancellation of construction plans. The author was lead litigator in the case.

MAGAZINES AND JOURNALS

Boustany, Katia. "The Development of Nuclear Law-Making or the Art of Legal 'Evasion'." *Nuclear Law Bulletin*, vol. 61, no. 1, June 1998, pp. 39–54. The author reviews the development of international laws dealing with nuclear power issues and observes how vague they tend to be, particularly in comparison with other international statutes and agreements, which tend to be highly detailed and specific.

Annotated Bibliography

Forinash, Betsy. "The US National System for Disposal of High-level and Transuranic Radioactive Wastes: Legislative History and Its Effect on Regulatory Approaches." *Nuclear Law Bulletin*, vol. 69, no. 1, June 2002, pp. 29–41. The author describes the development of policy that has led to the construction and planned construction of two waste disposal sites in the United States: the Waste Isolation Pilot Project (WIPP) in New Mexico and the proposed Yucca Mountain repository in Nevada.

INTERNET/WEB DOCUMENTS

U.S. Nuclear Regulatory Commission. "Nuclear Regulatory Legislation: 107th Congress; 1st Session (NUREG-0980, Vol. 1 & 2, Num. 6)." Available online. http://www.nrc.gov/reading-rm/doc-collections/nuregs/staff/sr0980/#publication_info. Downloaded on February 9, 2005. An invaluable source listing all federal legislation on nuclear energy issues passed up to and including the 1st session of the 107th Congress (2001–02).

CHAPTER 8

——————■——————

ORGANIZATIONS AND AGENCIES

This chapter contains information on agencies, associations, organizations, and other groups whose primary or exclusive focus involves some aspect of nuclear power. The list of organizations is divided into three general categories: (1) international, federal, and state agencies responsible for the regulation and (sometimes) promotion of nuclear power as a source of energy; (2) organizations interested in promoting the use of nuclear power and/or providing information about the use of nuclear power; and (3) groups promoting safety considerations in the use of nuclear power and/or opposed to the development of nuclear power plants.

INTERNATIONAL, FEDERAL, AND STATE AGENCIES

Many nations have nuclear programs that are administered, controlled, or supervised by some national agency. Space limitations make it impossible to list organizations in all those nations. Information on these organizations can often be found by searching the Internet for a category such as "atomic energy" or "nuclear energy," followed by the name of the country about which information is sought. This search will often lead to the homepages for regulatory agencies in other nations, such as the examples listed below.

NON-U.S. NATIONAL AGENCIES

Atomic Energy of Canada, Ltd. (AECL)
URL: http://www.aecl.ca
E-mail: info@aecl.ca
Phone: (905) 823-9040

2251 Speakman Drive
Mississauga, Ontario, L5K 1B2
Canada
Responsible for the promotion of nuclear power in Canada; roughly comparable to the U.S. Department of Energy, Office of Nuclear Energy, Science and Technology.

Canadian Coalition for Nuclear Responsibility (CCNR)
URL: http://www.ccnr.org
E-mail: ccnr@web.net
Phone: (514) 489-5118
C.P. 236, Station Snowdon
Montréal QC, H3X 3T4
Canada
CCNR was incorporated in 1978 as a not-for-profit federal organization dedicated to education and research on all issues related to nuclear energy, including both military and civilian applications, with particular attention to those issues pertaining to Canada.

Canadian Nuclear Safety Commission (CNSC)
URL: http://www.nuclearsafety.gc.ca
E-mail: info@cnsc-ccsn.gc.ca
Phone: (800) 668-5284 (in Canada) or (613) 995-5894 (outside Canada)
280 Slater Street
P.O. Box 1046, Station B
Ottawa, ON K1P 5S9
Canada
Responsible for regulation of the nuclear industry in Canada; roughly comparable to the U.S. Nuclear Regulatory Commission.

Commissariat à l'Energie Atomique/Siège (CEA)
URL: http://www.cea.fr
E-mail: webmaster@cea.fr
Phone: +33 (0)1 40 56 10 00
31-33 Rue de la Fédération
F-75752 Paris Cedex 15
France

Department of Atomic Energy (DAE)
URL: http://www.dae.gov.in
E-mail: webmanager@dae.gov.in
Phone: +91 (22)2202 6823 / 2202 8917 / 2202 8899 / 2286 2500
Government of India
Anushakti Bhavan
Chatrapathi Shivaji Maharaj Marg
Mumbai-400001
India

United Kingdom Atomic Energy Authority (UKAEA)
URL: http://www.ukaea.org.uk
E-mail: andrew.munn@ukaea.org
Phone: +44 (0)1235 820220
UKAEA Marshal Building
521 Downs Way
Harwell Dicot
Oxfordshire, OX11 ORA
United Kingdom

U.S. NATIONAL AGENCIES

Responsibility for research and development on nuclear power; the construction, licensing, and monitoring of nuclear power plants; and other nuclear-related issues is distributed within the U.S. government among a number of legislative and administrative departments, divisions, committees, and agencies. While the major responsibility in the U.S. Congress for such issues is assigned to the Senate Sub-committee on Science and the House Committee on Science, other committees and subcommittees in both houses are often involved in nuclear-related issues. For example, the Senate Committee on Environmental Quality has been very much involved in the problem of storing, transporting, and disposing of the nation's nuclear wastes.

The central agency for nuclear power in the administrative division of government is the U.S. Department of Energy (DOE), in general, and the Office of Nuclear Energy, Science and Technology, in particular. Again, many nuclear-related agencies are located in other divisions of the DOE and in other cabinet- and non-cabinet departments. For example, responsibility for dealing with the nation's nuclear waste disposal problem has been assigned primarily to the DOE's Office of Civilian Radioactive Waste Management (http://www.ocrwm.doe.gov), although a number of other DOE and non-DOE agencies are also involved in dealing with that issue. Also, most military applications of nuclear energy, along with some peacetime applications, for example, are located in the Department of Defense rather than the Department of Energy. And the Bureau of Land Management (BLM), in the Department of the Interior, has been very much involved in the construction of a nuclear waste dump at Yucca Mountain, Nevada, since the facility will be located on land administered by the BLM. Agencies and committees with responsibilities on nuclear issues of special interest in the United States are the following:

Energy Information Administration (EIA)
Department of Energy
URL: http://www.eia.doe.gov
E-mail: infoctr@eia.doe.gov
Phone: (202) 586-8800
1000 Independence Avenue, S.W., EI 30
Washington, DC 20585
The Energy Information Administration is a division of the U.S. Department of Energy whose responsibility it is to collect, collate, and make available to the public information on all aspects of energy production and use in the United States, including energy generated from nuclear power. One of the best and most reliable sources of data and statistics on energy in the United States.

National Council on Radiation Protection and Measurements (NCRP)
URL: http://www.ncrp.com
E-mail: jaszenko@ncrp.com
Phone: (301) 675-2652
7910 Woodmont Avenue
Suite 400
Bethesda, MD 20814-3095
The mission of the National Council on Radiation Protection and Measurement is to formulate

information and recommendations on radiation protection and measurement systems that represent the latest consensus of scientific thinking and to disseminate that information to professionals and the general public. This goal is accomplished primarily through the release of a number of reports, commentaries, symposia proceedings, statements, and other publications summarizing the organization's work.

National Nuclear Data Center (NNDC)
URL: http://www.nndc.bnl.gov
E-mail: nndc@bnl.gov
Phone: (631) 344-2902
Brookhaven National
  Laboratory
Building 197D
Upton, NY 11973-5000
NNDC collects, evaluates, and distributes technical information on a variety of topics in the area of nuclear energy, including nuclear structure and decay, nuclear reactions, bibliographic references, publications, and meetings. The majority of information provided on this site is highly technical, although some of the more general references may be of interest and value to the general reader.

National Nuclear Security Administration (NNSA)
URL: http://www.nnsa.doe.gov
E-mail: bpleau@doeal.gov
Phone: (505) 845-0011
NNSA Service Center

P. O. Box 5400
Albuquerque, NM 87185-5400
NNSA is responsible for enhancing the safety of the United States through the development of military nuclear weapons by ensuring the safety and effectiveness of such weapons and preventing the proliferation of nuclear weapons throughout the world. The agency's operations are carried out at a number of sites throughout the nation, including site offices at the Pittsburgh and Schenectady Naval Nuclear Reactors, the Nevada Test Site, Sandia National Laboratory, the Savannah River (South Carolina) Site, Lawrence Livermore National Laboratory, Los Alamos National Laboratory, the Pantex Plant (near Amarillo, Texas), the Honeywell Manufacturing Plant (in Kansas City, Missouri), and the Y-12 National Security Plant at the Oak Ridge National Laboratory.

Nuclear Regulatory Commission (NRC)
URL: http://www.nrc.gov
E-mail: opa@nrc.gov
Phone: (800) 368-5642
Washington, DC 20555-0001
(no street address required)
The Nuclear Regulatory Commission was established in the Energy Reorganization Act of 1974, as the ultimate successor to the Atomic Energy Commission's regulatory arm. The commission's mission is fourfold: to regulate the civilian use of nuclear materials, to ensure adequate protection of public health

and safety, to promote the common defense and security, and to protect the environment. NRC's responsibilities fall into three major areas: commercial reactors, used for the generation of electrical power, and research and test reactors, used for research and training; nuclear materials used in commercial reactors, in medical and industrial applications, and in research programs; and nuclear waste transportation, storage, and disposal, an area that includes decommissioning of nuclear power plants.

**Office of Civilian Radioactive
  Waste Management
  (OCRWM)**
URL: http://www.ocrwm.doe.gov
E-mail: info@ocrwm.doe.gov
Phone: (800) 225-6972
U. S. Department of Energy
**Office of Repository
  Development**
1551 Hillshire Drive
Las Vegas, NV 89134
The Office of Civilian Radioactive Waste Management was established within the Department of Energy as a provision of the Nuclear Waste Policy Act of 1982. It has primary responsibility for the nuclear waste repository being developed at Yucca Mountain, Nevada, and operates primarily out of two offices, one in Washington, D.C., and one in Nevada. The former office has responsibility for overall program administration, as well as scientific and technical problems relating to waste acceptance, storage, and transporta-

tion systems, while the latter office monitors scientific and engineering studies being conducted at Yucca Mountain to determine its suitability as a nuclear waste repository location.

**Office of Nuclear Energy,
  Science and Technology
  (ONEST)**
U.S. Department of Energy
URL: http://www.ne.doe.gov
**E-mail: ne.webmaster@hq.doe.
  gov**
Phone: (202) 586-6630
**1000 Independence Avenue,
  S.W.**
Washington, DC 20585-1290
ONEST is the primary governmental agency responsible for overall issues relating to nuclear energy. The agency has responsibilities in a number of specific areas, including nuclear facilities management, space and defense power systems, advanced nuclear research, and nuclear power systems.

**U.S. Department of Energy
  (DOE) National Laboratory
  System**
The U.S. Department of Energy National Laboratory System evolved out of the Manhattan Project during World War II, in which very large financial, personnel, and other resources were employed to work on the development of the first fission bombs for the U.S. government. Over time, those laboratories grew in number, size, and importance; today they account for nearly half of all the funding pro-

vided by the U.S. government for research in physics, chemistry, materials science, and other physical sciences. The mission of the national laboratories has also expanded and diversified from its original focus on the study of nuclear materials to a wide variety of research topics, including the development of new energy systems, environmental issues, basic biological and biomedical research, and human health projects. Of the 18 laboratories that make up the DOE National Laboratory System, five continue to have major nuclear-related projects. They are as follows:

**Brookhaven National
  Laboratory (BNL)**
URL: http://www.bnl.gov/world
E-mail: conrad@bnl.gov
Phone: (631) 344-8000
P.O. Box 5000
Upton, NY 11973-5000

Idaho National Laboratory (INL)
URL: http://www.inl.gov
E-mail: info@inel.gov
Phone: (800) 708-2680
2525 North Fremont Avenue
P.O. Box 1625
Idaho Falls, ID 83415

**Los Alamos National
  Laboratory (LANL)**
URL: http://www.lanl.gov/
  worldview
E-mail: community@lanl.gov
Phone: (505) 667-7000
P.O. Box 1663
Los Alamos, NM 87545

**Oak Ridge National Laboratory
  (ORNL)**
URL: http://www.ornl.gov
E-mail: hilldj@ornl.gov
Phone: (865) 574-4160
P.O. Box 2008
Oak Ridge, TN 37831

**Sandia National Laboratories
  (SNL)**
URL: http://www.sandia.gov
E-mail: webmaster@sandia.gov
Phone: (505) 284-5200
P.O. Box 5800
Albuquerque, NM 87185

**U.S. House Committee on
  Science**
URL: http://www.house.gov/
  science
E-mail: Science@mail.house.gov
Phone: (202) 226-6371
2320 Rayburn House Office
  Building
Washington, DC 20515
The House Committee on Science is the primary committee in the U.S. House of Representatives through which legislation relating to the peacetime applications of nuclear energy must pass. Agencies that fall under the committee's jurisdiction include the National Aeronautics and Space Administration (NASA), Department of Energy (DOE), Environmental Protection Agency (EPA), National Science Foundation (NSF), Federal Aviation Administration (FAA), National Oceanic and Atmospheric Administration (NOAA), National Institute of Standards and Technology

(NIST), Federal Emergency Management Agency (FEMA), U.S. Fire Administration, and U.S. Geological Survey.

U.S. Senate Subcommittee on Energy
URL: http://energy.senate.gov/about/about_energy.html
E-mail: energy-sub@energy.senate.gov
Phone: (202) 224-6567 (Majority); (202) 224-7571 (Minority)
Dirksen Senate Office Building 308 (Majority)
Dirksen Senate Office Building 312 (Minority)
Washington, DC 20510
The Subcommittee on Energy has responsibility for research and development of nuclear fuels, nuclear fuel policy, commercial nuclear fuel projects, nuclear fuels siting and insurance programs, and nuclear fuel cycle policy. It is the primary subcommittee in the U.S. Senate through which all peacetime applications of nuclear energy must pass.

INTERNATIONAL AGENCIES

In addition to national agencies such as those listed above, a number of international organizations exist to handle one or more nuclear power issues. Among the most important of these organizations are the following.

International Atomic Energy Agency (IAEA)
URL: http://www.iaea.or.at
E-mail: Official.Mail@iaea.org
Phone: +43 (1) 2600-0
P.O. Box 100
Wagramer Strasse 5
A-1400 Vienna
Austria
Established in 1957 as a response to the Atoms for Peace initiative, the IAEA is the world's primary agency for monitoring the peaceful uses of nuclear materials. In addition to carrying out inspections to ensure that nuclear materials are not being used for military purposes, the IAEA assists and advises nations on the development of peaceful applications of nuclear energy and the safe use of nuclear materials.

International Nuclear Energy Research Institute (INERI)
URL: http://www.pnl.gov/ineri
E-mail: richarem@id.doe.gov
Phone: (208) 526-2640
850 Energy Drive, MS 1221
Idaho Falls, ID 83401-1563
An agency created in 2001 by the U.S. Department of Energy for the purpose of promoting international cooperation in the development of the next generation of nuclear power plant technology. INERI has thus far developed bilateral agreements with Brazil, Canada, the European Union, France, the Organization for Economic Cooperation and Development—Nuclear Energy Agency, and the Republic of Korea for

programs on nuclear power plant development.

International Nuclear Law Association (INLA)
URL: http://www.aidn-inla.be
E-mail: info@aidn-inla.be
Phone: +32 (2) 547 58 41
Square de Meeûs 29
1000 Brussles
Belgium
INLA was organized in the 1970s as a way of bringing together legal scholars from around the world with special interest in problems and issues involving the peaceful applications of nuclear power. The organization sponsors lectures, conferences, seminars, congresses, and other meetings on specialized topics related to nuclear law.

International Nuclear Safety Center (INSC)
URL: http://www.insc.anl.gov
E-mail: inscdb@anl.gov
Phone: (630) 252-4713
Argonne National Laboratory
Building 208
9700 South Cass Avenue
Argonne, IL 60439
The INSC's mission is to promote research, collect and distribute information, and provide other services that will result in the improvement of nuclear reactor safety throughout the world. The center is under the overall guidance of the U.S. Department of Energy. It provides a variety of useful information for both specialists and the general public interested in nuclear power issues, information such as the location and characteristics of nuclear reactors around the world and an extensive summary of the properties of materials used in nuclear power facilities.

ORGANIZATIONS PROMOTING THE USE OF NUCLEAR POWER AND/OR PROVIDING INFORMATION ABOUT NUCLEAR POWER

American Institute of Physics (AIP)
URL: http://www.aip.org
E-mail: fyi@aip.org
Phone: (301) 209-3100
One Physics Ellipse
College Park, MD 20740-3843

AIP was founded in 1931 to promote the advancement of knowledge in physics, astronomy, and the related physical sciences and their application to human society. One of its primary functions is the publication of a number of books, journals, and other materials related to this mission. One of the society's most valuable resources for nuclear-related information is the Niels Bohr Library, which contains material on the history of the development of nuclear science (http://www.aip.org/history/nblbro.htm).

American Nuclear Society (ANS)
URL: http://www.ans.org
E-mail: Contact form available at www.ans.org/contact
Phone: (708) 352-6611
555 North Kensington Avenue
La Grange Park, IL 60526
An association of more than 10,000 engineers, scientists, administrators, and educators from more than 1,600 corporations, educational institutions, and government agencies interested in issues related to nuclear science.

Canadian Nuclear Association (CNA)
URL: http://www.cna.ca
E-mail: lemieuxc@cna.ca
Phone: (616) 237-4262
1610-130 Albert Street
Ottawa, ON K1P 5G4
Canada
The Canadian Nuclear Association was established as a nonprofit membership organization in 1960 to promote the role of nuclear power in Canada. The association seeks to create and foster a political environment and regulatory framework favorable to the use of nuclear energy in power production; to encourage cooperation among industries, educational institutions, governmental agencies, and other organizations involved in the production of nuclear power; to provide a forum for the discussion of problems related to the use of nuclear power in Canada; and to work in cooperation with other associations having similar goals and objectives.

Centre for Nuclear Energy Research (CNER)
URL: http://www.unb.ca/cner/web
E-mail: cner@unb.ca
Phone: (506) 453-5111
Incutech Building
Room 121
The University of New Brunswick
Fredericton, New Brunswick
E3B 6C2
Canada
CNER is a joint enterprise of the University of New Brunswick, the New Brunswick Research and Productivity Council, and Atomic Energy of Canada Limited. The center conducts research in the field of nuclear energy and is involved with the operation and maintenance of Canada's CANDU (Canada Deuterium Uranium) nuclear reactor.

Citizens for Medical Isotopes (CMI)
URL: http://www.medical isotopes.org
E-mail: info@medicalisotopes.org

Phone: (509) 737-8463
P.O. Box 802
Richland, WA 99352
CMI is an organization of physicians, research scientists, cancer patients and their families, and concerned citizens committed to educating the general public and legislators about the medical benefit of radioactive isotopes and promoting efforts to increase research and public expenditure on the development of such materials.

Division of Nuclear Physics (DNP)
American Physical Society (APS)
URL: http://dnp.nscl.msu.edu
E-mail: dnpweb@nscl.msu.edu
Phone: (301) 209-3200
One Physics Ellipse
College Park, MD 20740-3844
The Division of Nuclear Physics is one of 14 specialized departments within the American Physical Society. Members of the DNP are interested primarily in theoretical problems related to the structure of matter and its relationship to nuclear energy. The APS web site offers a particularly good general introduction to the topic of nuclear physics at http://www.aps.org/resources/nuclear.cfm.

Electric Power Research Institute (EPRI)
URL: http://www.epri.com
E-mail: chopf@epri.com
Phone: (650) 855-2733
3412 Hillview Avenue

P.O. Box 10412
Palo Alto, CA 94303
EPRI was created in 1971 by a consortium of private and public utility groups as a research institute aimed at solving fundamental problems in the generation and distribution of electrical energy in the United States. The organization has conducted an extensive array of projects on nuclear power, reports of which are available on its web site at http://www.epri.com/BMSprogram.asp?program=215955&targetlistyear=2004.

Environmental Literacy Council (ELC)
URL: http://www.enviroliteracy.org/subcategory.php/28.html
E-mail: info@enviroliteracy.org
Phone: (202) 296-0390
1625 K Street, N.W.
Suite 1020
Washington, DC 20006-3868
The Environmental Literacy Council is an independent nonprofit organization dedicated to the task of providing teachers with the curriculum tools they need to help students develop environmental literacy. It focuses on a number of specific topics, including air, land, water, energy, and ecosystems, of which nuclear energy is one subcategory. The council's web site provides detailed information on nuclear fission and fusion, nuclear security, and nuclear waste, especially the problems associated with the Yucca Mountain waste disposal program.

Environmentalists for Nuclear Energy (EFN)
URL: http://www.ecolo.org
E-mail: nuc-en@ecolo.org
Phone: +33 1 30 86 00 33 or
   +33 6 11 84 88 00
55 rue Victor Hugo
F-78800 Houilles
France
EFN was founded in 1996 to encourage anyone interested in environmental issues to support the development of nuclear power production as being one of the safest and least harmful of all methods of energy production currently available. Today the organization has about 5,000 members in 30 countries from all around the world. EFN sponsors annual meetings and encourages members and supporters to make use of a variety of techniques (such as letters, cards, telephone calls, e-mails) to influence decision makers' opinions about and actions on nuclear power production in their countries.

European Atomic Forum (FORATOM)
URL: http://www.foratom.org
E-mail: foratom@foratom.org
Phone: +32 (2) 502 45 95
Rue Belliard 15-17
1040 Brussels
Belgium
FORATOM is the trade association for nuclear power plant companies in Europe, with members accounting for about one-third of all electricity generated in the area. The association attempts to provide useful information to the general public; to decision makers, such as those in the European Parliament and the European Commission; and to the media. The organization also acts as a clearinghouse on information about nuclear power production among its member industries.

European Nuclear Society (ENS)
URL: http://www.euronuclear.
   org
E-mail: ens@euronuclear.org
Phone: +32 (2) 505 30 50
Rue Belliard 15-17
1040 Brussels
Belgium
The European Nuclear Society was founded in 1975 to promote the advancement of science and engineering in the peaceful application of nuclear power. The organization currently has 26 members from 25 countries, ranging from the United States and Canada to Israel and Russia. ENS offers nearly three dozen regular and special conferences and training sessions on a variety of technical topics every year.

Federation of American Scientists (FAS)
URL: http://www.fas.org
E-mail: webmaster@fas.org
Phone: (202) 546-3300
1717 K Street, N.W.
Suite 209
Washington, DC 20036
FAS was founded in 1945 by a group of scientists who had been

involved in the Manhattan Project for the production of the first nuclear weapons. These scientists were very concerned about the possible misuses of nuclear energy following World War II and created the organization to help provide guidance to the general public and decision makers as to the best uses of nuclear energy. Today FAS handles a variety of nuclear-related issues, ranging from weapons technology to the restoration of degraded land and the development of sustainable technologies.

Fusion Power Associates (FPA)
URL: http://ourworld.compu
serve.com/homepages/fpa
E-mail: fpa@compuserve.com
Phone: (301) 258-0545
2 Professional Drive
Suite 249
Gaithersburg, MD 20879
FPA is a nonprofit, tax-exempt research and educational foundation established to provide information on the status of fusion research and development and other applications of plasma science. The organization publishes bimonthly newsletters, sponsors annual management-level symposia, and distributes regular e-mail on developments in the field of fusion research.

Institute for Energy and
Environmental Research
(IEER)
URL: http://www.ieer.org

E-mail: ieer@ieer.org
Phone: (301) 270-5500
6935 Laurel Avenue
Suite 204
Takoma Park, MD 20912
The goal of IEER is to provide policy makers, journalists, and the general public with accurate information about important scientific and technological issues of the day, nuclear power among them. The organization hopes to promote the "democratization of science" so as to make possible a safer and healthier environment for all humans. The organization's web site provides a number of very useful "fabulous fact sheets" on a number of nuclear-related issues, including the basics of fissionable materials; the physical, chemical, and nuclear properties of plutonium; and nuclear waste disposal issues.

Institute of Nuclear Materials
Management (INMM)
URL: http://www.inmm.org
E-mail: inmm@inmm.org
Phone: (847) 480-9573
60 Revere Drive
Suite 500
Northbrook, IL 60062
INMM was created in 1958 to encourage the promotion of research in nuclear materials management, the establishment of standards for the field, the improvement of qualifications of those involved in nuclear materials management, and the dissemination of information about nuclear materials management through meetings, reports,

discussions, and publications. Membership is open to anyone involved in the development, teaching, and application of any aspect of nuclear materials management.

Institute of Nuclear Power Operations (INPO)
URL: http://www.eh.doe.gov/nsps/inpo
E-mail: esh-infocenter@eh.doe.gov
Phone: (770) 644-8000
700 Galleria Parkway
Atlanta, GA 30339
The Institute of Nuclear Power Operations is an organization within the U.S. Department of Energy consisting of all companies that operate nuclear power plants. The organization was established in 1979, following the Three Mile Island accident, to promote the development and use of safety and reliability in the operation of nuclear power plants.

Low Level Radiation Campaign (LLRC)
URL: http://www.llrc.org
E-mail: SiteManager@llrc.org
Phone: +44 (0) 1597 824771
The Knoll, Montpellier Park
Llandrindod Wells, Powys LD1 5LW
United Kingdom
LLRC explores scientific evidence relating to the human health effects of low levels of radiation from a wide variety of sources, including nuclear materials.

Managing the Atom
URL: http://bcsia.ksg.harvard.edu/research.cfm?program=STPP&project=MTA&pb_id=240&gma =27&gmi=47
E-mail: atom@harvard.edu
Phone: (617) 495-4219
Belfer Center for Science and International Affairs
John F. Kennedy School of Government
79 JFK Street
Cambridge, MA 02138
A project at the John F. Kennedy School of Government at Harvard University in which an international team of scholars and government officials study a range of issues relating to nuclear energy, ranging from weapons to nuclear power production.

National Atomic Museum (NAM)
URL: http://www.atomicmuseum.com
E-mail: info@atomicmuseum.com
Phone: (505) 245-2137
1905 Mountain Road NW
Albuquerque, NM 87104
The museum was established in 1969 to provide a resource on the history of the modern age of nuclear science. In addition to the exhibit space itself, the museum offers a Science Is Everywhere summer camp, traveling exhibits, and special educational programs.

North American Young Generation in Nuclear (NA-YGN)

URL: http://www.na-ygn.org
E-mail: naygn@na-ygn.org
Phone: (877) 526-2946
P.O. Box 10014
La Grange, IL 60525
NA-YGN is an organization of men and women under the age of 35 with a strong belief in and commitment to the development of nuclear power facilities. The organization sponsors a number of conferences and presentations annually, many of them in association with meetings of the American Nuclear Society. Its primary publication is a regular newsletter, *Go Nuke!*

Nuclear Energy Agency (NEA)
Organisation for Economic
 Co-operation and
 Development
URL: http://www.nea.fr
E-mail: nea@nea.fr
Phone: +33 (1) 45 24 10 10
Le Seine Saint-Germain
12, Boulevard des Îlles
F-92130 Issy-les-Moulineaux
France
An intergovernmental agency of 28 industrialized nations responsible for 85 percent of the developed nuclear power capacity in the world. The organization's purpose is to help member states develop and maintain "the scientific, technological and legal bases required for the safe, environmentally friendly and economical use of nuclear energy for peaceful purposes."

Nuclear Energy Institute (NEI)
URL: http://www.nei.org

E-mail: webmasterp@nei.org
Phone: (202) 739-8000
1776 I Street, N.W.
Suite 400
Washington, DC 20006-3708
An association consisting of nuclear utilities, plant designers, architectural and engineering firms, and fuel cycling companies whose purpose it is to promote the beneficial use of nuclear power in the United States and the world. NEI was established in 1994 through the merger of a number of other organizations, including the Atomic Industrial Forum, the Nuclear Utility Management and Resources Council, the U.S. Council for Energy Awareness, the American Nuclear Energy Council, and the nuclear division of the Edison Electric Institute. It currently has about 260 corporate members from 15 nations.

Nuclear Information and
 Resource Service & World
 Information Service on
 Energy (NIRS/WISE)
URL: http://www.nirs.org
E-mail: nirsnet@nirs.org
Phone: (202) 328-0002
1424 16th Street, N.W.
Suite 404
Washington, DC 20036
NIRS/WISE was formed in 1978 for the purpose of providing information to citizens and environmental organizations about nuclear power, nuclear waste disposal problems, radiation, and issues involving the sustainable use of energy. The organization sponsors a number of cam-

paigns on specific nuclear and energy-related issues, campaigns such as the NIRS Reactor Watchdog Project, the Don't Waste America program (on the Yucca Mountain waste disposal site), and the program on nuclear power and global warming.

Uranium Information Center (UIC)
URL: http://www.uic.com.au
E-mail: uicinfo@octopus.net.au
Phone: +61 (03) 9629 7744
GPO Box 1649
Melbourne 3001
Australia

UIC was founded in 1978 to provide information about the development of nuclear power facilities in Australia. The organization's web site has become, however, a treasure chest of information about every conceivable aspect of uranium and the process by which nuclear power is generated, along with issues related to that process, such as the problem of nuclear waste storage and disposal. Of particular value is its series of "briefing papers," all of which are available online at the organization's web site.

World Nuclear Association
URL: http://www.world-nuclear.org
E-mail: wna@world-nuclear.org
Phone: +44 (0)20 7225 0303
Bowater House West
12th floor
114 Knightsbridge
London SW1X 7LJ
United Kingdom

An association of individuals and companies interested in promoting the peaceful applications of nuclear energy throughout the world. WNA focuses on topics such as nuclear fuel production, economics of the nuclear industry, nuclear trade issues, radiological protection, transport of nuclear materials, and waste management and decommissioning.

ORGANIZATIONS PROMOTING SAFETY CONSIDERATIONS IN THE USE OF NUCLEAR POWER AND/OR OPPOSED TO NUCLEAR POWER PLANTS

Abalone Alliance Safe Energy Clearinghouse (AASEC)
URL: http://www.energy-net.org
E-mail: abalone@energy-net.org
Phone: (415) 861-0592
2940 16th Street, #310
San Francisco, CA 94103

An organization formed in May 1977 by 70 California activists opposed to the construction of a new nuclear power plant at Diablo

Canyon, near San Luis Obispo. The organization's web site has an excellent section on all aspects of the fundamentals of nuclear power production, a history of the Abalone Alliance, and information and contacts on various aspects of the safe production of energy by a variety of means.

Alliance for Nuclear Accountability (ANA)
URL: http://www.ananuclear.org
E-mail: jcbridgman@earthlink. net
Phone: (202) 544-0217
322 Fourth Street, N.E.
Washington, DC 20002
ANA is a network organization of 33 groups concerned with the health and environmental consequences of nuclear weapons, nuclear power production, and nuclear waste disposal. One of the organization's major events is an annual DC Days, at which activists from around the country come to Washington, D.C., to participate in training and advocacy activities.

Campaign for Nuclear Phaseout (CNP)
URL: http://www.cnp.ca/main
E-mail: cnp@web.net
Phone: (613) 789-3634
412-1 Nicholas Street
Ottawa, Ontario K1N 7B7
Canada
CNP is a coalition of Canadian public interest organizations concerned with the health and environmental consequences of nuclear

power plants and seeking to reduce and eventually eliminate the use of such plants in Canada. The organization's web site has a number of links to other Canadian groups opposed to the use of nuclear power plants.

Citizen Alert (CA)
URL: http://www.citizenalert. org
E-mail: pmj1@citizenalert.org
Phone: (702) 796-5662
P.O. Box 17173
Las Vegas, NV 89114
Citizen Alert was founded in 1975 in response to the federal government's plans to bury the nation's nuclear wastes at Yucca Mountain, Nevada. It is a network group consisting of more than a dozen independent, but cooperating, organizations. Citizen Alert has since expanded to include a number of other environmental issues within its purview and takes credit, for example, for preventing the deployment of the MX-missile in its area of the country.

Citizens Awareness Network (CAN)
URL: http://www.nukebusters. org
E-mail: can@nukebusters.org
Phone: (413) 339-5781
P.O. Box 83
Shelburne Falls, MA 01370
CAN was formed in 1991 when the owners of the Yankee Atomic reactor in Rowe, Massachusetts, announced their intention to apply for a renewal of its license with the

Nuclear Regulatory Commission. The organization eventually formed chapters in all the New England states to fight the continued operation of nuclear power plants in the region.

Committee for Nuclear Responsibility (CNR)
URL: http://www.ratical.org/ radiation/CNR
Phone: (415) 776-8299
P. O. Box 421993
San Francisco, CA 94142
CNR is a nonprofit educational organization founded in 1971 by John Gofman, a vigorous and vocal critic of government policy on the health effects of exposure to radiation. The purpose of the organization is to provide an independent viewpoint on the sources and health effects of ionizing radiation.

Critical Mass Energy and Environment Program (CMEEP)
URL: http://www.citizen.org/ cmep
E-mail: CMEP@citizen.org
Phone: (202) 588-1000
1600 20th Street, N.W.
Washington, DC 20009
CMEEP is a program of the organization Public Citizen. Its goal is to protect citizens and the environment from dangers posed by nuclear power plants and to promote efforts to find safer, more affordable, and more environmentally friendly methods of generating power.

Energy Probe
URL: http://www.energyprobe. org
E-mail: TomAdams@nextcity. com
Phone: (416) 964-9223
225 Brunswick Avenue
Toronto, Ontario M5S 2M6
Canada
Energy Probe is a consumer and environment research team dedicated to working against the continuation and extension of nuclear power plant construction and operation. The organization's most successful efforts have been in forcing the clean-up of sites contaminated by nuclear operations and improving the regulatory oversight of existing plants.

Greenpeace
URL: http://www.greenpeace usa.org
E-mail: info@wdc.greenpeace.org
Phone: (800) 326-0959
702 H Street, N.W.
Suite 300
Washington, DC 20001
One of the oldest, largest, and most highly respected environmental organizations, Greenpeace is very active in the field of nuclear energy, actively working in opposition to the construction of new nuclear power plants and the relicensing of existing facilities while searching for a way of dealing with the nation's nuclear waste problems.

Indian Point Safe Energy Coalition (IPSEC)
URL: http://www.ipsecinfo.org

E-mail: ipsecpc@bestweb.ne
Phone: (888) 474-8848
P.O. Box 134
Croton-on-Hudson, NY 10520
IPSEC was formed shortly after the terrorist attacks of September 11, 2001, in response to widespread concerns about the vulnerability of the Indian Point nuclear power plant, in Buchanan, New York, to future attacks by terrorists. The organization now includes about 65 environmental, health, and public policy groups within its network. IPSEC's activities consist largely of lobbying governmental agencies to bring about closing the plant.

Military Toxics Project (MTP)
URL: http://www.miltoxproj.org
E-mail: mtp@miltoxproj.org
Phone: (207) 783-5091
P.O. Box 558
Lewiston, ME 04243-0558
Although focusing in particular on issues of nuclear wastes produced by military activities, MTP is also concerned more generally with questions of the production, transportation, processing, storage, and disposal of nuclear wastes for a variety of sources. Three of the organization's major projects deal with contaminated military bases, depleted uranium materials, and environmental problems created by conventional munitions production, storage, and use.

National Resources Defense Council (NRDC)
URL: http://www.nrdc.org

E-mail: nrdcinfo@nrdc.org
Phone: (212) 727-2700
40 West 20th Street
New York, NY 10011
NRDC is an organization of more than a million members dedicated to trying to save the Earth's environment. One of the organization's major campaigns focuses on Nuclear Weapons & Waste, for which it provides position papers, research news, policy statements, and other resources on its web page at http://www.nrdc.org/nuclear/default.asp.

Nuclear Control Institute (NCI)
URL: http://www.nci.org
E-mail: nci@nci.org
Phone: (202) 822-8444
1000 Connecticut Avenue, N.W.
Suite 410
Washington, DC 20036
NCI was founded in 1981 as a research and advocacy organization designed primarily to prevent the proliferation of nuclear weapons and nuclear materials throughout the world. Although the organization's primary emphasis is on nuclear weapons, it also conducts research and provides information on issues related to nuclear power production, such as the release of reports on the transport, processing, storage, and disposal of nuclear wastes.

Nuclear Energy Information Service (NEIS)
URL: http://www.neis.org
E-mail: neis@neis.org

Phone: (847) 869-7650
P.O. Box 1637
Evanston, IL 60204-1637
An organization committed to ending the use of nuclear power. NEIS efforts include educating and organizing the general public on energy issues; building grassroots, nonviolent opposition to nuclear power; and advocating sustainable and ecologically sound energy alternatives.

**Nuclear Power Research
    Institute (NPRI)**
URL: http://www.nuclearpolicy.
    org
E-mail: info@nuclearpolicy.org
Phone: (202) 822-9800
1925 K Street, N.W.
Suite 210
Washington, DC 20006
NPRI was founded in 2002 by Dr. Helen Caldicott, who also founded Physicians for Social Responsibility. The organization's goal is to educate the general public about the threats posed to the environment and human health by nuclear weapons, nuclear power production, and nuclear wastes.

Nukewatch
URL: http://www.nukewatch.
    com
E-mail: nukewatch@lakeland.ws
Phone: (715) 472-4185
P.O. Box 649
Luck, WI 54853
Nukewatch was organized in 1979 in response to concerns about the secrecy surrounding the develop-
ment of nuclear weapons and nuclear power plant production facilities. The organization's web site provides a number of interesting documents dealing with a number of nuclear-related issues, including radioactive waste and transport, food irradiation, depleted uranium materials, nuclear materials in space, and nuclear power production.

Sierra Club
URL: http://www.sierraclub.org
E-mail: information@sierraclub.
    org
Phone: (415) 977-5500
85 Second Street
Second Floor
San Francisco, CA 94105
The Sierra Club claims to be "the oldest, largest, and most influential" environmental advocacy group in the United States, with interests in a broad range of issues. Its concerns with nuclear power relate primarily to the problem of nuclear waste storage, transportation, and disposal, topics discussed in detail on the organization's webpage at http://www.sierraclub.org/nuclearwaste.

Tri-Valley CAREs
URL: http://www.trivalleycares.
    org
E-mail: marylia@earthlink.net
Phone: (925) 443-7148
2582 Old First Street
Livermore, CA 94551
Tri-Valley CAREs (Communities Against a Radioactive Environment) is an organization formed in 1983

by citizens concerned about the potential environmental effects of two national laboratories, Lawrence Livermore and Sandia. The organization has worked to have the laboratories shut down, cleaned up, and/or converted to other, less dangerous types of research.

Union of Concerned Scientists (UCS)
URL: http://www.ucsusa.org
E-mail: ssi@ucsusa.org
Phone: (617) 547-5552
2 Brattle Square
Cambridge, MA 02238-9105
UCS was founded by a group of faculty members and students at the Massachusetts Institute of Technology in 1969, concerned about the real and possible misuses of science in modern society. The organization's activities in the field of nu-

clear power are described on its Web page at https://www.ucsusa. org/clean_energy/nuclear_safety/ index.cfm.

Wisconsin's Nuclear WatchDog (WNWD)
URL: http://www.wnwd.org
E-mail: mail@psrmadison.org
Phone: (608) 232-9945
PSR Madison
P.O. Box 1712
Madison, WI 53701-1712
A subsidiary of Physicians for Social Responsibility, WNWD focuses on nuclear issues such as power generation, safety, waste, regulation, economics, and weapons in the state of Wisconsin. The organization's web site also provides a place for the collection and distribution of information about nuclear science.

PART III

APPENDICES

APPENDIX A

ATOMIC ENERGY ACT OF 1946 (PUBLIC LAW 79-585)

[The Atomic Energy Act of 1946 was the first piece of legislation adopted by the U.S. Congress that dealt with the control of nuclear energy in the United States. The following excerpt omits certain sections that discuss the creation of the Atomic Energy Commission (Section 2), Patents (Section 10), Enforcement (Section 12), Reports (Section 13), Definitions (Section 14), Appropriations (Section 15), Separability of Provisions (Section 16), and Short Title (Section 17), as well as certain portions within sections, as noted.]

DECLARATION OF POLICY

Section 1. (a) Findings and Declaration—Research and experimentation in the field of nuclear fission have attained the stage at which the release of atomic energy on a large scale is practical. The significance of the atomic bomb for military purposes is evident. The effect of the use of atomic energy for civilian purposes upon the social, economic, and political structures of today cannot now be determined. It is reasonable to anticipate, however, that tapping this new source of energy will cause profound changes in our present way of life. Accordingly, it is hereby declared to be the policy of the people of the United States that the development and utilization of atomic energy shall be directed toward improving the public welfare, increasing the standard of living, strengthening free competition among private enterprises so far as practicable, and cementing world peace.

(b) Purpose of Act.—It is the purpose of this Act to effectuate these policies by providing, among others, for the following major programs:

(1) A program of assisting and fostering private research and development on a truly independent basis to encourage maximum scientific progress;

(2) A program for the free dissemination of basic scientific information and for maximum liberality in dissemination of related technical information;

(3) A program of federally conducted research to assure the Government of adequate scientific and technical accomplishments;

(4) A program for Government control of the production, ownership, and use of fissionable materials to protect the national security and to insure the broadest possible exploitation of the field;

(5) A program for simultaneous study of the social, political, and economic effects of the utilization of atomic energy; and

(6) A program of administration which will be consistent with international agreements made by the United States, and which will enable the Congress to be currently informed so as to take further legislative action as may hereafter be appropriate. . . .

RESEARCH

Sec. 3. (a) Research Assistance.—The Commission is directed to exercise its powers in such a manner as to insure the continued conduct of research and developmental activities in the fields specified below by private or public institutions or persons and to assist in the acquisition of an ever-expanding fund of theoretical and practical knowledge in such fields. To this end the Commission is authorized and directed to make contracts, agreements, arrangements, grants-in-aid, and loans—

(1) for the conduct of research and developmental activities relating to (a) nuclear processes; (b) the theory and production of atomic energy, including processes and devices related to such production; (c) utilization of fissionable and radioactive materials for medical or health purposes; (d) utilization of fissionable and radioactive materials for all other purposes, including industrial uses; and (e) the protection of health during research and production activities; and

(2) for studies of the social, political, and economic effects of the availability and utilization of atomic energy. . . .

PRODUCTION OF FISSIONABLE MATERIALS

Sec. 4. (a) Definition.—The term "production of fissionable materials" shall include all methods of manufacturing, producing, refining, or processing fissionable materials, including the process of separating fissionable material from other substances in which such material may be contained, whether by thermal diffusion, electromagnetic separation, or other processes.

(b) Authority to Produce.—The Commission shall be the exclusive producer of fissionable materials, except production incident to research or developmental activities subject to restrictions provided in subparagraph (d) below. The quantities of fissionable material to be produced in any quarter shall be determined by the President.

(c) Prohibition.—It shall be unlawful for any person to produce any fissionable material except as may be incident to the conduct of research or developmental activities.

(d) Research and Development on Production Processes.

(1) The Commission shall establish by regulation such requirements for the reporting of research and developmental activities on the production of fissionable materials as will assure the Commission of full knowledge of all activities, rates of production, and quantities produced.

(2) The Commission shall provide for the frequent inspection of all such activities by employees of the Commission.

(3) No person may in the course of such research or developmental activities possess or operate facilities for the production of fissionable material in quantities or at a rate sufficient to construct a bomb or other military weapon unless all such facilities are the property of and subject to the control of the Commission. The Commission is authorized, to the extent that it deems such action consistent with the purposes of this Act, to enter into contracts for the conduct of such research or developmental activities involving the use of the Commission's facilities. . . .

CONTROL OF MATERIALS

[Section 5 restricts the ownership of fissionable materials to the U.S. government, except under those circumstances under which the Atomic Energy Commission may provide those materials to persons or organizations authorized to use those materials for specific, approved purposes.]

MILITARY APPLICATIONS OF ATOMIC POWER

Sec. 6. (a) The Commission is authorized and directed to —

(1) conduct experiments and do research and developmental work in the military application of atomic power; and

(2) have custody of all assembled or unassembled atomic bombs, bomb parts, or other atomic military weapons, presently or hereafter produced, except that upon the express finding of the President that such action is required in the interests of national defense, the Commission shall deliver such quantities of weapons to the armed forces as the President may specify.

(b) The Commission shall not conduct any research or developmental work in the military application of atomic power if such research or developmental work is contrary to any international agreement of the United States.

(c) The Commission is authorized to engage in the production of atomic bombs, bomb parts, or other applications of atomic power as military weapons, only to the extent that the express consent and direction of the President of the United States has been obtained, which consent and direction shall be obtained for each quarter.

(d) It shall be unlawful for any person to manufacture, produce, or process any device or equipment designed to utilize fissionable materials as a military weapon, except as authorized by the Commission.

ATOMIC ENERGY DEVICES

Sec. 7. (a) License Required.—It shall be unlawful for any person to operate any equipment or device utilizing fissionable materials without a license issued by the Commission authorizing such operation.

(b) Issuance of Licenses.—Any person desiring to utilize fissionable materials in any such device or equipment shall apply for a license therefor in accordance with such procedures as the Commission may by regulation establish. The Commission is authorized and directed to issue such a license on a nonexclusive basis and to supply appropriate quantities of fissionable materials to the extent available to any applicant (1) who is equipped to observe such safety standards to protect health and to minimize danger from explosion as the Commission may establish; and (2) who agrees to make available to the Commission such technical information and data concerning the operation of such device as the Commission may determine necessary to encourage the use of such devices by as many licensees as possible. Where any license might serve to maintain or foster the growth of monopoly, restraint of trade, unlawful competition, or other trade position inimical to the entry of new, freely competitive enterprises, the Commission is authorized and directed to refuse to issue such license or to establish such conditions to prevent these results as the Commission, in consultation with the Attorney General, may determine. The Commission shall report promptly to the Attorney General any information it may have of the use of such devices which appears to have these results. No license may be given to a foreign government or to any person who is not under and within the jurisdiction of the United States.

(c) Byproduct Power.—If in the production of fissionable materials the production processes yield energy capable of utilization, such energy may be used by the Commission, transferred to other Government agencies, sold to public or private utilities under contract providing for reasonable resale prices, or sold to private consumers at reasonable rates and on as broad a basis of eligibility as the Commission may determine to be possible.

[Section (d) requires the Atomic Energy Commission to make reports to the Congress whenever "in its opinion industrial, commercial, or other uses of fissionable materials have been sufficiently developed to be of practical value."]

PROPERTY OF THE COMMISSION

Sec. 8. (a) The President shall direct the transfer to the Commission of the following property owned by the United States or any of its agencies, or any interest in such property held in trust for or on behalf of the United States:

(1) All fissionable materials; all bombs and bomb parts; all plants, facilities, equipment, and materials for the processing or production of fissionable materials, bombs, and bomb parts; all processes and technical information of any kind, and the source thereof (including data, drawings, specifications, patents, patent applications, and other sources, relating to the refining or production of fissionable materials; and all contracts, agreements, leases, patents, applications for patents, inventions and discoveries (whether patented or unpatented), and other rights of any kind concerning any such items;

(2) All facilities and equipment, and materials therein, devoted primarily to atomic energy research and development; and

(3) All property in the custody and control of the Manhattan engineer district.

(b) In order to render financial assistance to those States and local governments in which the activities of the Commission are carried on and in which the Commission, or its agents, have acquired properties previously subject to State and local taxation, the Commission is authorized to make payments to State and local governments in lieu of such taxes. Such payments may be in the amounts, at the times, and upon the terms the Commission deems appropriate, but the Commission shall be guided by the policy of not exceeding the taxes which would have been payable for such property in the condition in which it was acquired, except where special burdens have been cast upon the State or local government by activities of the Commission, the Manhattan engineer district, or their agents, and in such cases any benefits accruing to the States and local governments by reason of these activities shall be considered in the determination of such payments. The Commission and any corporation created by it, and the property and income of the Commission or of such corporation, are hereby expressly exempted from taxation in any manner or form by any State, county, municipality, or any subdivision thereof.

DISSEMINATION OF INFORMATION

Sec. 9. (a) Basic Scientific Information.—Basic scientific information in the fields specified in section 3 may be freely disseminated. The term "basic scientific information" shall include, in addition to theoretical knowledge of nuclear and other physics, chemistry, biology, and therapy, all results capable of accomplishment, as distinguished from the processes or techniques of accomplishing them.

(b) Related Technical Information.—The Commission shall establish a Board of Commission. The Board shall, under the direction and supervision of the Commission, provide for the dissemination of related technical information with the utmost liberality as freely as may be consistent with the foreign and domestic policies established by the President and shall have authority to —

(1) establish such information services, publications, libraries, and other registers of available information as may be helpful in effectuating this policy;

(2) designate by regulation the types of related technical information the dissemination of which will effectuate the foregoing policy. Such designations shall constitute an administrative determination that such information is not of value to the national defense and that any person is entitled to receive such information, within the meaning of the Espionage Act. Failure to make any such designation shall not, however, be deemed a determination that such undesignated information is subject to the provisions of said Act;

(3) by regulation or order, require reports of the conduct of independent research or development activities in the fields specified in section 3 and of the operation of atomic energy devices under licenses issued pursuant to section 7;

(4) provide for such inspections of independent research and development activities of the types specified in section 3 and of the operation of atomic energy devices as the Commission or the Board may determine; and

(5) whenever it will facilitate the carrying out of the purposes of the Act, adopt by regulation administrative interpretations of the Espionage Act ex-

cept that any such interpretation shall, before adoption, receive the express approval of the President. . . .

ORGANIZATIONAL AND GENERAL AUTHORITY

Sec. 11. (a) Organization.—There are hereby established within the Commission a Division of Research, a Division of Production, a Division of Materials, and a Division of Military Application. Each division shall be under the direction of a Directory who shall be appointed by the President, by and with the advice and consent of the Senate, and shall receive compensation at the rate of $15,000 per annum. The Commission shall delegate to each such division such of its powers under this Act as in its opinion from time to time will promote the effectuation of the purposes of this Act in an efficient manner. Nothing in this paragraph shall prevent the Commission from establishing such additional divisions or other subordinate organizations as it may deem desirable.

(b) General Authority.—In the performance of its functions the Commission is authorized to —

(1) establish advisory boards to advise with and make recommendations to the Commission on legislation, policies, administration, and research;

(2) establish by regulation or order such standards and instructions to govern the possession and use of fissionable and byproduct materials as the Commission may deem necessary or desirable to protect health or to minimize danger from explosion;

(3) make such studies and investigations, obtain such information, and hold such hearings as the Commission may deem necessary or proper to assist it in exercising any authority provided in this Act, or in the administration or enforcement of this Act, or any regulations or orders issued thereunder. For such purposes the Commission is authorized to require any person to permit the inspection and copying of any records or other documents, to administer oaths and affirmations, and by subpena [sic] to require any person to appear and testify, or to appear and produce documents, or both, at any designated place. Witnesses subpenaed [sic] under this subsection shall be paid the same fees and mileage as are paid witnesses in the district courts of the United States;

(4) create or organize corporations, the stock of which shall be wholly owned by the United States and controlled by the Commission, to carry out the provisions of this Act; . . .

APPENDIX B

———

ATOMIC ENERGY ACT OF 1954 (PUBLIC LAW 83-703)

[The Atomic Energy Act of 1954 enhanced and updated the Atomic Energy Act of 1946 and was much longer (157 pages, with later additions and changes) than its predecessor. The excerpts below include some of the most important provisions of the 1954 act, some of which were later modified or deleted by later amendments. The complete act can be accessed on the Nuclear Regulatory Commission's homepage at http://www.nrc.gov/who-we-are/governing-laws.html.]

TITLE I—ATOMIC ENERGY

CHAPTER 1—DECLARATION, FINDINGS, AND PURPOSE

Sec. 1. Declaration

Atomic energy is capable of application for peaceful as well as military purposes. It is therefore declared to be the policy of the United States that—

a. the development, use, and control of atomic energy shall be directed so as to make the maximum contribution to the general welfare, subject at all times to the paramount objective of making the maximum contribution to the common defense and security; and

b. the development, use, and control of atomic energy shall be directed so as to promote world peace, improve the general welfare, increase the standard of living, and strengthen free competition in private enterprise.

Sec. 2. Findings.

The Congress of the United States hereby makes the following findings concerning the development, use, and control of atomic energy:

a. The development, utilization, and control of atomic energy for military and for all other purposes are vital to the common defense and security.

b. In permitting the property of the United States to be used by others such sue [*probably should be* use] must be regulated in the national interest and in order to provide for the common defense and security and to protect the health and safety of the public. [*deleted by Public Law 88-489*]

c. The processing and utilization of source, byproduct, and special nuclear material affect interstate and foreign commerce and must be regulated in the national interest.

d. The processing and utilization of source, byproduct, and special nuclear material must be regulated in the national interest and in order to provide for the common defense and security and to protect the health and safety of the public.

e. Source and special nuclear material, production facilities, and utilization facilities are affected with the public interest, and regulation by the United States of the production and utilization of atomic energy and of the facilities used in connection therewith is necessary in the national interest to assure [sic] the common defense and security and to protect the health and safety of the public.

f. The necessity for protection against possible interstate damage occurring from the operation of facilities for the production or utilization of source or special nuclear material places the operation of those facilities in interstate commerce for the purposes of this Act.

g. Funds of the United States may be provided for the development and use of atomic energy under conditions which will provide for the common defense and security and promote the general welfare.

h. It is essential to the common defense and security that title to all special nuclear material be in the United States while such special nuclear material is within the United States.

i. In order to protect the public and to encourage the development of the atomic energy industry, in the interest of the general welfare and of the common defense and security, the United States may make funds available for a portion of the damages suffered by the public from nuclear incidents, and may limit the liability of those persons liable for such losses.

Sec. 3. Purpose.

It is the purpose of this Act to effectuate the policies set forth above by providing for—

a. a program of conducting, assisting, and fostering research and development in order to encourage maximum scientific and industrial progress;

b. a program for the dissemination of unclassified scientific and technical information and for the control, dissemination, and declassification of Restricted Data, subject to appropriate safeguards, so as to encourage scientific and industrial progress;

c. a program for Government control of the possession, use, and production of atomic energy and special nuclear material, whether owned by the Government or others, so directed as to make the maximum contribution to the common defense and security and the national welfare, and to provide continued assurance of the Government's ability to enter into and enforce agreements with nations or groups of nations for the control of special nuclear materials and atomic weapons.

d. a program to encourage widespread participation in the development and utilization of atomic energy for peaceful purposes to the maximum extent consistent with the common defense and security and with the health and safety of the public;

e. a program of international cooperation to promote the common defense and security and to make available to cooperating nations the benefits of peaceful applications of atomic energy as widely as expanding technology and considerations of the common defense and security will permit; and

f. a program of administration which will be consistent with the foregoing policies and programs, with international arrangements, and with agreements for cooperation, which will enable the Congress to be currently informed so as to take further legislative action as may be appropriate. . . .

CHAPTER 4—RESEARCH

Sec. 31. Research Assistance.

a. The Commission is directed to exercise its powers in such manner as to insure the continued conduct of research and development and training activities in the fields specified below, by private or public institutions or persons, and to assist in the acquisition of an ever-expanding fund of theoretical and practical knowledge in such fields. To this end the Commission is authorized and directed to make arrangements (including contracts, agreements, and loans) for the conduct of research and development activities relating to —

(1) nuclear processes;

(2) the theory and production of atomic energy, including processes, materials, and devices related to such production;

(3) utilization of special nuclear material and radioactive material for medical, biological, agricultural, health, or military purposes;

(4) utilization of special nuclear material, atomic energy, and radioactive material and processes entailed in the utilization or production of atomic energy or such material for all other purposes, including industrial or commercial uses, the generation of usable energy, and the demonstration of advances in the commercial or industrial application of atomic energy;

(5) the protection of health and the promotion of safety during research and production activities; and

(6) the preservation and enhancement of a viable environment by developing more efficient methods to meet the Nation's energy needs.

b. The Commission is further authorized to make grants and contributions to the cost of construction and operation of reactors and other facilities and other equipment to colleges, universities, hospitals, and eleemosynary [philanthropic] or charitable institutions for the conduct of educational and training activities relating to the fields in subsection a.

c. The Commission may (1) make arrangements pursuant to this section, without regard to the provisions of section 3709 of the Revised Statutes, as amended, upon certification by the Commission that such action is necessary in the interest of the common defense and security, or upon a showing by the Commission that advertising is not reasonably practicable; (2) make partial and advance payments under such arrangements; and (3) make available for use in connection therewith such of its equipment and facilities as it may deem desirable.

d. The arrangements made pursuant to this section shall contain such provisions (1) to protect health, (2) to minimize danger to life or property, and (3) to require the reporting and to permit the inspection of work performed thereunder, as the Commission may determine. No such arrangement shall contain any provisions or conditions which prevent the dissemination of scientific or technical information, except to the extent such dissemination is prohibited by law.

Sec. 32 Research by the Commission.

The Commission is authorized and directed to conduct, through its own facilities, activities and studies of the types specified in section 31.

Sec. 33 Research for Others.

Where the Commission finds private facilities or laboratories are inadequate for the purpose, it is authorized to conduct for other persons, through its own facilities, such of those activities and studies of the types specified in section 31 as it deems appropriate to the development of energy. To the extent the Commission determines that private facilities or laboratories are inadequate for the purpose, and that the Commission's facilities, or scientific or technical resources have the potential of lending significant assistance to other persons in the fields of protection of public health and safety, the Commission may also assist other persons in these fields by conducting for such persons, through the Commission's own facilities, research and development or training activities and studies. The Commission is authorized to determine and make such charges as in its discretion may be desirable for the conduct of the activities and studies referred to in this section.

CHAPTER 5—PRODUCTION OF
SPECIAL NUCLEAR MATERIAL

Sec. 41. Ownership and Operation of Production Facilities.

a. Ownership of Production Facilities—The Commission, as agent of and on behalf of the United States, shall be the exclusive owner of all production facilities other than facilities which (1) are useful in the conduct of research and development activities in the fields specified in section 31, and do not, in the opinion of the Commission, have a potential production rate adequate to enable the user of such facilities to produce within a reasonable period of time a sufficient quantity of special nuclear material to produce an atomic weapon; (2) are licensed by the Commission under this title; or (3) are owned by the United States Enrichment Corporation.

b. Operation of the Commission's Facilities—The Commission is authorized and directed to produce or to provide for the production of special nuclear material in its own production facilities. To the extent deemed necessary, the Commission is authorized to make, or to continue in effect, contracts with persons obligating them to produce special nuclear material in facilities owned by the Commission. The Commission is also authorized to enter into research and development contracts authorizing the contractor to produce special nuclear material in facilities owned by the Commission to the extent that the production of such special nuclear material may be incident to the conduct of research and development activities under such contracts. Any contract entered into under this section shall contain provisions (1) prohibiting the contractor from subcontracting any part of the work he is obligated to perform under the contract, except as authorized by the Commission; and (2) obligating the contractor (A) to make such reports pertaining to activities under the contract to the Commission as the Commission may require, (B) to submit to inspection by employees of the Commission of all such activities, and (C) to comply with all safety and security regulations which may be prescribed by the Commission. Any contract made under the provisions of this subsection may be made without regard to the provisions of section 3079 of the Revised Statutes, as amended, upon certification by the Commission that such action is necessary in the interest of the common defense and security, or upon a showing by the Commission that advertising is not reasonably practicable. Partial and advance payments may be made under such contracts.

c. Operation of Other Production Facilities—Special nuclear material may be produced in the facilities which under this section are not required to be owned by the Commission. . . .

Appendix B

CHAPTER 10—ATOMIC ENERGY LICENSES

Sec. 101. License Required.

It shall be unlawful, except as provided in section 91, for any person within the United States to transfer or receive in interstate commerce, manufacture, produce, transfer, acquire, possess, use, import, or export any utilization or production facility except under and in accordance with a license issued by the Commission pursuant to section 103 or 104. . . .

Sec. 103. Commercial Licenses.

a. The Commission is authorized to issue licenses to persons applying therefor to transfer or receive in interstate commerce, manufacture, produce, transfer, acquire, possess, use, import, or export under the terms of an agreement for cooperation arranged pursuant to section 123, utilization or production facilities for industrial or commercial purposes. Such licenses shall be issued in accordance with the provisions of chapter 16 and subject to such conditions as the Commission may by rule or regulation establish to effectuate the purposes and provisions of this Act.

b. The Commission shall issue such licenses on a nonexclusive basis to persons applying therefor (1) whose proposed activities will serve a useful purpose proportionate to the quantities of special nuclear material or source material to be utilized; (2) who are equipped to observe and who agree to observe such safety standards to protect health and to minimize danger to life or property as the Commission may by rule establish; and (3) who agree to make available to the Commission such technical information and data concerning activities under such licenses as the Commission may determine necessary to promote the common defense and security and to protect the health and safety of the public. All such information may be used by the Commission only for the purposes of the common defense and security and to protect the health and safety of the public.

c. Each such license shall be issued for a specified period, as determined by the Commission, depending on the type of activity to be licensed, but not exceeding forty years, and may be renewed upon the expiration of such period.

d. No license under this section may be given to any person for activities which are not under or within the jurisdiction of the United States, except for the export of production or utilization facilities under terms of an agreement for cooperation arranged pursuant to section 123, or except under the provisions of section 109. No license may be issued to an alien or any corporation or other entity if the Commission knows or has reason to believe it is owned, controlled, or dominated by an alien, a foreign corporation, or a foreign government. In any event, no license may be issued to any person within the United States if, in the opinion of the Commission, the issuance of a license to such person would be inimical to the common defense and security or to the health and safety of the public.

f. Each license issued for a utilization facility under this section or section 104(b) shall require as a condition thereof that in case of any accident which could result in an unplanned release of quantities of fission products in excess of allowable limits for normal operation established by the Commission, the licensee shall immediately so notify the Commission. Violation of the condition prescribed by this subsection may, in the Commission's discretion, constitute grounds for license revocation. In accordance with section 187, the Commission shall promptly amend each license for a utilization facility issued under this section or section 104(b) which is in effect on the date of enactment of this subsection to include the provisions required under this subsection.

Sec. 104. Medical Therapy and Research and Development.

a. The Commission is authorized to issue licenses to persons applying therefor for utilization facilities for use in medical therapy. In issuing such licenses the Commission is directed to permit the widest amount of effective medical therapy possible with the amount of special nuclear material available for such purposes and to impose the minimum amount of regulation consistent with its obligations under this chapter to promote the common defense and security and to protect the health and safety of the public.

b. As provided for in subsection 102b, or 102c, or where specifically authorized by law, the Commission is authorized to issue licenses under this subsection to persons applying therefor for utilization and production facilities for industrial and commercial purposes. In issuing licenses under this subsection, the Commission shall impose the minimum amount of such regulations and terms of license as will permit the Commission to fulfill its obligations under this Act.

c. The Commission is authorized to issue licenses to persons applying therefor for utilization and production facilities useful in the conduct of research and development activities of the types specified in section 31 and which are not facilities of the type specified in subsection 104b of this section. The Commission is directed to impose only such minimum amount of regulation of the licensee as the Commission finds will permit the Commission to fulfill its obligations under this chapter to promote the common defense and security and to protect the health and safety of the public and will permit the conduct of widespread and diverse research and development.

d. No license under this section may be given to any person for activities which are not under or within the jurisdiction of the United States, except for the export of production or utilization facilities under terms of an agreement for cooperation arranged pursuant to section 109 or except under the provisions of section 109. No license may be issued to any corporation or other entity if the Commission knows or has reason to believe it is owned, controlled, or dominated by an alien, a foreign corporation, or a foreign government. In any event, no license may be issued to any person within the United States if, in the opinion of the Commission, the issuance of a license to such person would be inimical to the common defense and security or to the health and safety of the public. . . .

CHAPTER 16—JUDICIAL REVIEW AND ADMINISTRATIVE PROCEDURE

Sec. 181. General.

The provisions of the Administrative Procedure Act (Public Law 404, Seventy-ninth Congress, approved June 11, 1946) shall apply to all agency action taken under this Act, and the terms 'agency' and 'agency action'; shall have the meaning specified in the Administrative Procedure Act: Provided, however, that in the case of agency proceedings or actions which involve Restricted Data or defense information, the Commission shall provide by regulation for such parallel procedures as will effectively safeguard and prevent disclosure of Restricted Data or defense information to unauthorized persons with minimum impairment of the procedural rights which would be available if Restricted Data or defense information were not involved.

Sec. 182. License Applications.

a. Each application for a license hereunder shall be in writing and shall specifically state such information as the Commission, by rule or regulation, may determine to be necessary to decide such of the technical and financial qualifications of the applicant, the character of the applicant, the citizenship of the applicant, or any other qualifications of the applicant as the Commission may deem appropriate for the license. In connection with applications for licenses to operate production or utilization facilities, the applicant shall state such technical specifications, including information of the amount, kind, and source of special nuclear material required, the place of the use, the specific characteristics of the facility, and such other information as the Commission may, by rule or regulation, deem necessary in order to enable it to find that the utilization, or production of special nuclear material will be in accord with the common defense and security and will provide adequate protection to the health and safety of the public. Such technical specifications shall be a part of any license issued. The Commission may at any time after the filing of the original application, and before the expiration of the license, require further written statements in order to enable the Commission to determine whether the application should be granted or denied or whether a license should be modified or revoked. All applications and statements shall be signed by the applicant or licensee under oath or affirmation. . . .

c. The Commission shall not issue any license for a utilization or production facility for the generation of commercial power under section 103, until it has given notice in writing to such regulatory agency as may have jurisdiction over the rates and services of the proposed activity, to municipalities, private utilities, public bodies, and cooperatives within transmission distance authorized to engage in the distribution of electric energy and until it has published notice of such application once each week for four consecutive weeks in the Federal Register, and until four weeks after the last notice.

d. The Commission, in issuing any license for utilization or production facility for the generation of commercial power under section 103, shall give preferred consideration to applications for such facilities which will be located in high cost power areas in the United States if there are conflicting applications resulting from limited opportunity for such license. Where such conflicting applications resulting from limited opportunity for such license include those submitted by public or cooperative bodies such applications shall be given preferred consideration.

Sec. 183. Terms of Licenses.

Each license shall be in such form and contain such terms and conditions as the Commission may, by rule or regulation, prescribe to effectuate the provisions of this Act, including the following provisions:

a. Title to all special nuclear material utilized or produced by facilities pursuant to the license shall at all times be in the United States.

b. No right to the special nuclear material shall be conferred by the license except as defined by the license.

c. Neither the license nor any right under the license shall be assigned or otherwise transferred in violation of the provisions of this Act.

d. Every license issued under this Act shall be subject to the right of recapture or control reserved by section 108, and to all of the other provisions of this Act, now or hereafter in effect and to all valid rules and regulations of the Commission.

Sec. 184. Inalienability of Licenses.

No license granted hereunder and no right to utilize or produce special nuclear material granted hereby shall be transferred, assigned or in any manner disposed of, either voluntarily or involuntarily, directly or indirectly, through transfer of control of any license to any person, unless the Commission shall, after securing full information, find that the transfer is in accordance with the provisions of this Act, and shall give its consent in writing. The Commission may give such consent to the creation of a mortgage, pledge, or other lien upon any facility owned or thereafter acquired by a licensee, or upon any leasehold or other interest in such property, and the rights of the creditors so secured may thereafter be enforced by any court subject to rules and regulations established by the Commission to protect public health and safety and promote the common defense and security.

Sec. 185. Construction Permits.

All applicants for licenses to construct or modify production or utilization facilities shall, if the application is otherwise acceptable to the Commission, be initially granted a construction permit. The construction permit shall state the earliest and latest dates for the completion of the construction or modification. Unless the construction or modification of the facility is completed by the completion date, the construction permit shall expire, and all rights thereunder be forfeited, unless upon good cause shown, the Commission extends the completion date. Upon the completion of the construction or modification of the facility, upon the filing of any additional information needed to bring the original application up to date, and upon finding that the facility authorized has been constructed and will operate in conformity with the application as amended and in conformity with the provisions of this Act and of the rules and regulations of the Commission, and in the absence of any good cause being shown to the Commission why the granting of a license would not be in accordance with the provisions of this Act, the Commission shall thereupon issue a license to the applicant. For all other purposes of this Act, a construction permit is deemed to be a "license.". . .

Sec. 186. Revocation.

a. Any license may be revoked for any material false statement in the application or any statement of fact required under section 182, or because of conditions revealed by such application or statement of fact or any report, record, or inspection or other means which would warrant the Commission to refuse to grant a license on an original application, or for failure to construct or operate a facility in accordance with the terms of the construction permit or license or the technical specifications in the application, or for violation of, or failure to observe any of the terms and provisions of this Act or of any regulation of the Commission.

b. The Commission shall follow the provisions of section 9(b) of the Administrative Procedure Act in revoking any license.

c. Upon revocation of the license, the Commission may immediately retake possession of all special nuclear material held by the licensee. In cases found by the Commission to be of extreme importance to the national defense and security or to the health and safety of the public, the Commission may recapture any special nuclear material held by the licensee or may enter upon and operate the facility prior to any of the procedures provided under the Administrative Procedure Act. Just compensation shall be paid for the use of the facility. . . .

APPENDIX C

NUCLEAR WASTE POLICY ACT OF 1982 (PUBLIC LAW 97-425)

[In the nearly four decades following the beginning of the modern atomic age in the 1940s, the U.S. government largely ignored the problem of nuclear waste disposal. During that time, thousands of tons of high- and low-level radioactive wastes had been stored "temporarily," usually at the locations where they had been produced. In 1982, the U.S. Congress passed the Nuclear Waste Policy Act, in which it laid out a general philosophy for the storage of nuclear wastes and a plan by which that philosophy could be implemented. The current law (P.L. 97-425) contains both the elements of the original act and amendments that have been passed, most important, those relating to the selection of Yucca Mountain, Nevada, as the site at which storage is supposed to take place.]

TITLE I—DISPOSAL AND STORAGE OF HIGH-LEVEL RADIOACTIVE WASTE, SPENT NUCLEAR FUEL, AND LOW-LEVEL RADIOACTIVE WASTE STATE AND AFFECTED INDIAN TRIBE PARTICIPATION IN DEVELOPMENT OF PROPOSED REPOSITORIES FOR DEFENSE WASTE

Sec. 101. . . . (a) Notification to States and affected Indian tribes. Notwithstanding the provisions of section 8 [42 U.S.C. 10107], upon any decision by the Secretary or the President to develop a repository for the disposal of high-level radioactive waste or spent nuclear fuel resulting exclusively from atomic energy defense activities, research and development activities of the Secretary, or both, and before proceeding with any site-specific investigations with respect to such repository, the Secretary shall notify the Governor and legislature of the State in which such repository is proposed to be located, or the governing body of the affected Indian tribe on whose reservation such repository is proposed to be located, as the case may be, of such decision.

(b) Participation of States and affected Indian tribes. Following the receipt of any notification under subsection (a), the State or Indian tribe involved shall be entitled, with respect to the proposed repository involved, to rights of participation and consultation identical to those provided in sections 115 through 118 [42 U.S.C. 10135–10138], except that any financial assistance authorized to be provided to such State or affected Indian tribe under section 116(c) or 118(b) [42 U.S.C. 10136(c), 10138(b)] shall be made from amounts appropriated to the Secretary for purposes of carrying out this section. [42 U.S.C. 10121]

SUBTITLE A—REPOSITORIES FOR DISPOSAL OF HIGH-LEVEL RADIOACTIVE WASTE AND SPENT NUCLEAR FUEL FINDINGS AND PURPOSES

Sec. 111. . . . (a) Findings. The Congress finds that—

(1) radioactive waste creates potential risks and requires safe and environmentally acceptable methods of disposal;

(2) a national problem has been created by the accumulation of (A) spent nuclear fuel from nuclear reactors; and (B) radioactive waste from (i) reprocessing of spent nuclear fuel; (ii) activities related to medical research, diagnosis, and treatment; and (iii) other sources;

(3) Federal efforts during the past 30 years to devise a permanent solution to the problems of civilian radioactive waste disposal have not been adequate;

(4) while the Federal Government has the responsibility to provide for the permanent disposal of high-level radioactive waste and such spent nuclear fuel as may be disposed of in order to protect the public health and safety and the environment, the costs of such disposal should be the responsibility of the generators and owners of such waste and spent fuel;

(5) the generators and owners of high-level radioactive waste and spent nuclear fuel have the primary responsibility to provide for, and the responsibility to pay the costs of, the interim storage of such waste and spent fuel until such waste and spent fuel is accepted by the Secretary of Energy in accordance with the provisions of this Act [42 U.S.C. 10101 et seq.];

(6) State and public participation in the planning and development of repositories is essential in order to promote public confidence in the safety of disposal of such waste and spent fuel; and

(7) high-level radioactive waste and spent nuclear fuel have become major subjects of public concern, and appropriate precautions must be taken to ensure that such waste and spent fuel do not adversely affect the public health and safety and the environment for this or future generations.

(b) Purposes. The purposes of this subtitle [42 U.S.C. 10131 et seq.] are—

(1) to establish a schedule for the siting, construction, and operation of repositories that will provide a reasonable assurance that the public and the environment will be adequately protected from the hazards posed by high-level radioactive waste and such spent nuclear fuel as may be disposed of in a repository;

(2) to establish the Federal responsibility, and a definite Federal policy, for the disposal of such waste and spent fuel;

(3) to define the relationship between the Federal Government and the State governments with respect to the disposal of such waste and spent fuel; and

(4) to establish a Nuclear Waste Fund, composed of payments made by the generators and owners of such waste and spent fuel, that will ensure that the costs of carrying out activities relating to the disposal of such waste and spent fuel will be borne by the persons responsible for generating such waste and spent fuel. [42 U.S.C. 10131]

Sec. 112. Recommendation of Candidate Sites for Site Characterization

(a) Guidelines. Not later than 180 days after the date of the enactment of this Act [enacted Jan. 7, 1983], the Secretary, following consultation with the Council on Environmental Quality, the Administrator of the Environmental Protection Agency, the Director of the Geological Survey, and interested Governors, and the concurrence of the Commission shall issue general guidelines for the recommendation of sites for repositories. Such guidelines shall specify detailed geologic considerations that shall be primary criteria for the selection of sites in various geologic media. Such guidelines shall specify factors that qualify or disqualify any site from development as a repository, including factors pertaining to the location of valuable natural resources, hydrology, geophysics, seismic activity, and atomic energy defense activities, proximity to water supplies, proximity to populations, the effect upon the rights of users of water, and proximity to components of the National Park System, the National Wildlife Refuge System, the National Wild and Scenic Rivers System, the National Wilderness Preservation System, or National Forest Lands. Such guidelines shall take into consideration the proximity to sites where high-level radioactive waste and spent nuclear fuel is generated or temporarily stored and the transportation and safety factors involved in moving such waste to a repository. Such guidelines shall specify population factors that will disqualify any site from development as a repository if any surface facility of such repository would be located (1) in a highly populated area; or (2) adjacent to an area 1 mile by 1 mile having a population of not less than 1,000 individuals. Such guidelines also shall require the Secretary to consider the cost and impact of transporting to the repository site the solidified high-level radioactive waste and spent fuel to be disposed of in the repository and the advantages of regional distribution in the siting of repositories. Such guidelines shall require the Secretary to consider the various geologic media in which sites for repositories may be located and, to the extent practicable, to recommend sites in different geologic media. The Secretary shall use guidelines established under this subsection in considering candidate sites for recommendation under subsection (b). The Secretary may revise such guidelines from time to time, consistent with the provisions of this subsection. . . .

Definitions

[On May 27, 1986, President Ronald Reagan selected Yucca Mountain in Nevada as the site for storage of nuclear wastes. The decision resulted in a number of changes to the original 1982 legislation, including the following items.]

Sec. 2. For purposes of this Act [42 U.S.C. 10101 et seq.]:

(30) The term "Yucca Mountain site" means the candidate site in the State of Nevada recommended by the Secretary to the President under section 112(b)(1)(B) [42 U.S.C. 10132(b)(1)(B)] on May 27, 1986.

Sec. 113. Site Characterization

(a) In general. The Secretary shall carry out, in accordance with the provisions of this section, appropriate site characterization activities at the Yucca Mountain site. The Secretary shall consider fully the comments received under subsection (b)(2) and section 112(b)(2) [42 U.S.C. 10132(b)(2)] and shall, to the maximum extent practicable and in consultation with the Governor of the State of Nevada, conduct site characterization activities in a manner that minimizes any significant adverse environmental impacts identified in such comments or in the environmental assessment submitted under section 112(b)(1).

[Twenty four pages of text, constituting 13 additional sections, follow the above introduction, detailing the steps to be taken in preparing the Yucca Mountain site for the acceptance of nuclear wastes.]

Nuclear Power

SUBTITLE B—INTERIM STORAGE PROGRAM FINDINGS AND PURPOSES

Sec. 131. . . (a) Findings. The Congress finds that—

(1) the persons owning and operating civilian nuclear power reactors have the primary responsibility for providing interim storage of spent nuclear fuel from such reactors, by maximizing, to the extent practical, the effective use of existing storage facilities at the site of each civilian nuclear power reactor, and by adding new onsite storage capacity in a timely manner where practical;

(2) the Federal Government has the responsibility to encourage and expedite the effective use of existing storage facilities and the addition of needed new storage capacity at the site of each civilian nuclear power reactor; and

(3) the Federal Government has the responsibility to provide, in accordance with the provisions of this subtitle, not more than 1,900 metric tons of capacity for interim storage of spent nuclear fuel for civilian nuclear power reactors that cannot reasonably provide adequate storage capacity at the sites of such reactors when needed to assure the continued, orderly operation of such reactors.

(b) Purposes. The purposes of this subtitle [42 U.S.C 10151 et seq.] are—

(1) to provide for the utilization of available spent nuclear fuel pools at the site of each civilian nuclear power reactor to the extent practical and the addition of new spent nuclear fuel storage capacity where practical at the site of such reactor; and

(2) to provide, in accordance with the provisions of this subtitle [42 U.S.C. 10151 et seq.], for the establishment of a federally owned and operated system for the interim storage of spent nuclear fuel at one or more facilities owned by the Federal Government with not more than 1,900 metric tons of capacity to prevent disruptions in the orderly operation of any civilian nuclear power reactor that cannot reasonably provide adequate spent nuclear fuel storage capacity at the site of such reactor when needed. [42 U.S.C. 10151]

Sec. 132. Available Capacity for Interim Storage of Spent Nuclear Fuel

The Secretary, the Commission, and other authorized Federal officials shall each take such actions as such official considers necessary to encourage and expedite the effective use of available storage, and necessary additional storage, at the site of each civilian nuclear power reactor consistent with—

(1) the protection of the public health and safety, and the environment;
(2) economic considerations;
(3) continued operation of such reactor;
(4) any applicable provisions of law; and
(5) the views of the population surrounding such reactor. . . .

Appendix C

TITLE II—RESEARCH, DEVELOPMENT, AND DEMONSTRATION REGARDING DISPOSAL OF HIGH-LEVEL RADIOACTIVE WASTE AND SPENT NUCLEAR FUEL

[Title II of the Act contains the core instructions by which a disposal site for high-level radioactive wastes is to be identified, studied, tested, and evaluated. Sections 211 through 225 deal with a number of related topics, such as Siting Research and Related Activities (Sec. 214), Research and Development on Disposal of High-Level Radioactive Wastes (Sec. 217), Payments to States and Indian Tribes (Sec. 219), Judicial Review (Sec. 221), Subseabed Disposal (Sec. 224), and Dry Cask Storage (Sec. 225). The overall purpose of the title is outlined in Section 211, as follows.]

Sec. 211. Purpose
It is the purpose of this title [42 U.S.C. 10191 et seq.]—

    (1) to provide direction to the Secretary with respect to the disposal of high-level radioactive waste and spent nuclear fuel;

    (2) to authorize the Secretary, pursuant to this title [42 U.S.C. 10191 et seq.]—

        (A) to provide for the construction, operation, and maintenance of a deep geologic test and evaluation facility; and

        (B) to provide for a focused and integrated high-level radioactive waste and spent nuclear fuel research and development program, including the development of a test and evaluation facility to carry out research and provide an integrated demonstration of the technology for deep geologic disposal of high-level radioactive waste, and the development of the facilities to demonstrate dry storage of spent nuclear fuel; and

    (3) to provide for an improved cooperative role between the Federal Government and States, affected Indian tribes, and units of general local government in the siting of a test and evaluation facility.

[42 U.S.C. 10191] . . .

APPENDIX D

U.S. SUPREME COURT:
BALTIMORE GAS & ELECTRIC CO. v. NRDC, 462 U.S. 87 (1983)

[For many critics of nuclear power, the potential for environmental damage as a result of nuclear waste storage has long been an important issue. In this case, the U.S. Supreme Court ruled on the fundamental questions raised by that concern. The opening paragraph of Justice Sandra Day O'Connor's opinion outlines the issues presented by the case and the Court's position on that issue. (Citations and footnotes are omitted from the following extract.) The full decision can be read online at http://caselaw.lp.findlaw.com/cgi-bin/getcase.pl?navby=volpage&court=us&vol=462&page=100.]

Section 102(2)(C) of the National Environmental Policy Act (NEPA) requires federal agencies to consider the environmental impact of any major federal action. The dispute in these cases concerns the adoption by the Nuclear Regulatory Commission (NRC) of a series of generic rules to evaluate the environmental effects of a nuclear powerplant's fuel cycle. In these rules, the NRC decided that licensing boards should assume, for purposes of NEPA, that the permanent storage of certain nuclear wastes would have no significant environmental impact (the so-called "zero-release" assumption) and thus should not affect the decision whether to license a particular nuclear powerplant. At the heart of each rule is Table S-3, a numerical compilation of the estimated resources used and effluents released by fuel cycle activities supporting a year's operation of a typical light-water reactor. Challenges to the rules ultimately resulted in a decision by the Court of Appeals, on a petition for review of the final version of the rules, that the rules were arbitrary and capricious and inconsistent with NEPA because the NRC had not factored the consideration of uncertainties surrounding the zero-release assumption into the licensing process in such a manner that the uncertainties could potentially affect the outcome of any decision to license a plant.

Held:

The NRC complied with NEPA, and its decision is not arbitrary or capricious within the meaning of 10(e) of the Administrative Procedure Act (APA). Pp. 97–108.

[The Court briefly reviews the environmental issues involved in the nuclear fuel cycle and discusses the recent legal history surrounding challenges to the way in which the Nuclear Regulatory Commission has dealt with these issues.]

I

The environmental impact of operating a light-water nuclear powerplant includes the effects of offsite activities necessary to provide fuel for the plant ("front end" activities), and of offsite activities necessary to dispose of the highly toxic and long-lived nuclear wastes generated by the plant ("back end" activities). The dispute in these cases concerns the Commission's adoption of a series of generic rules to evaluate the environmental effects of a nuclear powerplant's fuel cycle. At the heart of each rule is Table S-3, a numerical compilation of the estimated resources used and effluents released by fuel cycle activities supporting a year's operation of a typical light-water reactor. The three versions of Table S-3 contained similar numerical values, although the supporting documentation has been amplified during the course of the proceedings.

[Here the Court reviews the challenges raised by the Natural Resources Defense Council (NRDC) and other organizations to NRC's past and current policy on assessing the environmental effects of "front end" and "back end" activities. The Court then discusses its views on the arbitrariness of the "zero-release" assumption.]

II

We are acutely aware that the extent to which this Nation should rely on nuclear power as a source of energy is an important and sensitive issue. Much of the debate focuses on whether development of nuclear generation facilities should proceed in the face of uncertainties about their long-term effects on the environment. Resolution of these fundamental policy questions lies, however, with Congress and the agencies to which Congress had delegated authority, as well as with state legislatures and, ultimately, the populace as a whole. Congress has assigned the courts only the limited, albeit important, task of reviewing agency action to determine whether the agency conformed with controlling statutes. As we emphasized in our earlier encounter with these very proceedings, "[a]dminstrative decisions should be set aside in this context, as in every other, only for substantial procedural or substantive reasons as mandated by statute . . . not simply because the court is unhappy with the result reached." [Vermont Yankee, *435 U.S., at 558.*]

The controlling statute at issue here is NEPA. NEPA has twin aims. First, it "places upon an agency the obligation to consider every significant aspect of the environmental impact of a proposed action." Vermont Yankee. Second, it ensures that the agency will inform the public that it has indeed considered environmental concerns in its decisionmaking process. Congress in enacting NEPA, however, did not require agencies to elevate environmental concerns over other appropriate considerations. Rather, it required only that the agency take a "hard look" at the environmental consequences before taking a major action. The role of the courts is simply to ensure that the agency has adequately considered and disclosed the environmental impact of its actions and that its decision is not arbitrary or capricious.

In its Table S-3 rule here, the Commission has determined that the probabilities favor the zero-release assumption, because the Nation is likely to develop methods to store the wastes with no leakage to the environment. The NRDC did not challenge and the Court of Appeals did not decide the reasonableness of this determination, and no party seriously challenges it here. The Commission recognized, however, that the geological, chemical, physical, and other data it relied on in making this prediction were based, in part, on assumptions which involve substantial uncertainties. Again, no one suggests that the uncertainties are trivial or the potential effects insignificant if time proves the zero-release assumption to have been seriously wrong. After confronting the issue, though, the Commission has determined that the uncertainties concerning the development of nuclear waste storage facilities are not sufficient to affect the outcome of any individual licensing decision.

It is clear that the Commission, in making this determination, has made the careful consideration and disclosure required by NEPA. The sheer volume of proceedings before the Commission is impressive. Of far greater importance, the Commission's Statement of Consideration announcing the final Table S-3 rule shows that it has digested this mass of material and disclosed all substantial risks. The Statement summarizes the major uncertainty of long-term storage in bedded-salt repositories, which is that water could infiltrate the repository as a result of such diverse factors as geologic faulting, a meteor strike, or accidental or deliberate intrusion by man. The Commission noted that the probability of intrusion was small, and that the plasticity of salt would tend to heal some types of intrusions. The Commission also found the evidence "tentative but favorable" that an appropriate site could be found. Table S-3 refers interested persons to staff studies that discuss the uncertainties in greater detail. Given this record and the Commission's statement, it simply cannot be said that the Commission ignored or failed to disclose the uncertainties surrounding its zero-release assumption.

Congress did not enact NEPA, of course, so that an agency would contemplate the environmental impact of an action as an abstract exercise. Rather, Congress intended that the "hard look" be incorporated as part of the agency's process of deciding whether to pursue a particular federal action. It was on this ground that the Court of Appeals faulted the Commission's action, for failing to allow the uncertainties potentially to "tip the balance" in a particular licensing decision. As a general proposition, we can agree with the Court of Appeals' determination that an agency must allow all significant environmental risks to be factored into the decision whether to undertake a proposed action. We think, however, that the Court of Appeals erred in concluding that the Commission had not complied with this standard.

As Vermont Yankee made clear, NEPA does not require agencies to adopt any particular internal decisionmaking structure. Here, the agency has chosen to evaluate generically the environmental impact of the fuel cycle and inform individual licensing boards, through the Table S-3 rule, of its evaluation. The generic method chosen by the agency is clearly an appropriate method of conducting the "hard look" required by NEPA. The environmental effects of much of the fuel cycle are not plant specific, for any plant, regardless of its particular attributes, will create additional wastes that must be stored in a common long-term repository. Administrative efficiency and consistency of decision are both furthered by a generic determination of these effects without needless repetition of the litigation in individual proceedings, which are subject to review by the Commission in any event.

The Court of Appeals recognized that the Commission has discretion to evaluate generically the environmental effects of the fuel cycle and require that these values be "plugged into" individual licensing decisions. The court concluded that the Commission nevertheless violated NEPA by failing to factor the uncertainty surrounding long-term storage into Table S-3 and precluding individual licensing decisionmakers from considering it.

The Commission's decision to affix a zero value to the environmental impact of long-term storage would violate NEPA, however, only if the Commission acted arbitrarily and capriciously in deciding generically that the uncertainty was insufficient to affect any individual licensing decision. In assessing whether the Commission's decision is arbitrary and capricious, it is crucial to place the zero-release assumption in context. Three factors are particularly important. First is the Commission's repeated emphasis that the zero-release assumption—and, indeed, all of the Table S-3 rule—was made for a limited purpose. The Commission expressly noted its intention to supplement the rule with an explanatory narrative. It also emphasized that the purpose of the rule was not to evaluate or select the most effective long-term waste disposal technology or develop site selection criteria. A separate and comprehensive series of programs has been undertaken to serve these broader purposes. In the proceedings before us, the Commission's staff did not attempt to evaluate the environmental effects of all possible methods of disposing of waste. Rather, it chose to analyze intensively the most probable long-term waste disposal method—burial in a bedded-salt repository several hundred meters below ground—and then "estimate its impacts conservatively, based on the best available information and analysis." The zero-release assumption cannot be evaluated in isolation. Rather, it must be assessed in relation to the limited purpose for which the Commission made the assumption.

Second, the Commission emphasized that the zero-release assumption is but a single figure in an entire Table, which the Commission expressly designed as a risk-averse estimate of the environmental impact of the fuel cycle. It noted that Table S-3 assumed that the fuel storage canisters and the fuel rod cladding would be corroded before a repository is closed and that all volatile materials in the fuel would escape to the environment. Given that assumption, and the improbability that materials would escape after sealing, the Commission determined that the overall Table represented a conservative (i.e., inflated) statement of environmental impacts. It is not unreasonable for the Commission to counteract the uncertainties in postsealing releases by balancing them with an overestimate of presealing releases. A reviewing court should not magnify a single line item beyond its significance as only part of a larger Table.

Third, a reviewing court must remember that the Commission is making predictions, within its area of special expertise, at the frontiers of science. When examining this kind of scientific determination, as opposed to simple findings of fact, a reviewing court must generally be at its most deferential.

With these three guides in mind, we find the Commission's zero-release assumption to be within the bounds of reasoned decisionmaking required by the APA. We have already noted that the Commission's Statement of Consideration detailed several areas of uncertainty and discussed why they were insubstantial for purposes of an individual licensing decision. The Table S-3 rule also refers to the staff reports, public documents that contain a more expanded discussion of the uncertainties involved in concluding that long-term storage will have no environmental effects. These staff reports recognize that rigorous verification of long-term risks for waste repositories is not possible, but suggest that data and extrapolation of past experience allow the Commission to identify events that could produce repository failure, estimate the probability of those events, and calculate the resulting consequences. The Commission staff also modeled the consequences of repository failure by tracing the flow of contaminated water, and found them to be insignificant. Ultimately, the staff concluded that

> *"[t]he radiotoxic hazard index analyses and the modeling studies that have been done indicate that consequences of all but the most improbable events will be small. Risks (probabilities times consequences) inherent in the long term for geological disposal will therefore also be small."*

We also find significant the separate views of Commissioners Bradford and Gilinsky. These Commissioners expressed dissatisfaction with the zero-release assumption and yet emphasized the limited purpose of the assumption and the overall conservatism of Table S-3. Commissioner Bradford characterized the bedded-salt repository as a responsible working assumption for NEPA purposes and concurred in the zero-release figure because it does not appear to affect Table S-3's overall conservatism. Commissioner Gilinsky was more critical of the entire Table, stating that the Commission should confront directly whether it should license any nuclear reactors in light of the problems of waste disposal, rather than hide an affirmative conclusion to this issue behind a table of numbers. He emphasized that the "waste confidence proceeding" should provide the Commission an appropriate vehicle for a thorough evaluation of the problems involved in the Government's commitment to a waste disposal solution. For the limited purpose of individual licensing proceedings, however, Commissioner Gilinsky found it "virtually inconceivable" that the Table should affect the decision whether to license, and characterized as "naive" the notion that the fuel cycle effluents could tip the balance in some cases and not in others.

In sum, we think that the zero-release assumption—a policy judgment concerning one line in a conservative Table designed for the limited purpose of individual licensing decisions—is within the bounds of reasoned decisionmaking. It is not our task to determine what decision we, as Commissioners, would have reached. Our only task is to determine whether the Commission has considered the relevant factors and articulated a rational connection between the facts found and the choice made. Under this standard, we think the Commission's zero-release assumption, within the context of Table S-3 as a whole, was not arbitrary and capricious.

[Part III deals in more detail with the validity of the so-called Table S-3 in making legitimate decisions about nuclear power plant licensing. The Court concludes that:]

In short, we find it totally inappropriate to cast doubt on licensing proceedings simply because of a minor ambiguity in the language of the earlier rule under which the environmental impact statement was made, when there is no evidence that this ambiguity prevented any party from making as full a presentation as desired, or ever affected the decision to license the plant. . . .

APPENDIX E

U.S. SUPREME COURT: *SILKWOOD V. KERR-MCGEE CORP.*, 464 U.S. 238 (1984)

[This case reached the U.S. Supreme Court after an Oklahoma jury had awarded Silkwood actual damages in the amount of $505,000 ($500,000 for personal injuries and $5,000 for property damage), and punitive damages of $10 million, a decision that was later reversed by the Tenth Circuit Court of Appeals. Justice Byron White delivered the Court's decision in this case on January 11, 1984. The full decision is available online on the FindLaw web site at http://caselaw.lp.findlaw. com/scripts/getcase.pl?court=us&vol=464&invol=238. Footnotes and citations are omitted in the following extract.]

Last Term, this Court examined the relationship between federal and state authority in the nuclear energy field and concluded that States are precluded from regulating the safety aspects of nuclear energy. *Pacific Gas & Electric Co. v. State Energy Resources Conservation & Development Comm'n*, 461 U.S. 190 (1983). This case requires us to determine whether a state-authorized award of punitive damages arising out of the escape of plutonium from a federally licensed nuclear facility is pre-empted either because it falls within that forbidden field or because it conflicts with some other aspect of the Atomic Energy Act.

[In Section I of the Court's opinion, omitted here, Justice White describes in some detail the history of the case, beginning with Karen Silkwood's exposure to radioactive materials while an employee at Kerr-McGee, her decision to sue the company for wrongful injury, and the basis on which the two lower courts reached their decisions. In Section II, also omitted here, he discusses the jurisdictional issues involved in the case. In Section III, White explains the Court's own position in the matter:]

III

As we recently observed in *Pacific Gas & Electric Co. v. State Energy Resources Conservation & Development Comm'n* (1983), state law can be pre-empted in either of two general ways. If Congress evidences an intent to occupy a given field, any state law falling within that field is pre-empted. If Congress has not entirely displaced state regulation over the matter in question, state law is still pre-empted to the extent it actually conflicts with federal law, that is, when it is impossible to comply with both state and federal law, or where the state law stands as an obstacle to the accomplishment of the full purposes and objectives of Congress. Kerr-McGee contends that the award in this case is invalid under either analysis. We consider each of these contentions in turn.

A

In *Pacific Gas & Electric*, an examination of the statutory scheme and legislative history of the Atomic Energy Act convinced us that "Congress . . . intended that the Federal Government should regulate the radiological safety aspects involved in the construction and operation of a nuclear plant." Thus, we concluded that "the Federal Government has occupied the entire field of nuclear safety concerns, except the limited powers expressly ceded to the States."

Kerr-McGee argues that our ruling in *Pacific Gas & Electric* is dispositive of the issue in this case. Noting that "regulation can be as effectively exerted through an award of damages as through some form of preventive relief," Kerr-McGee submits that because the state-authorized award of punitive damages in this case punishes and deters conduct related to radiation hazards, it falls within the prohibited field. However, a review of the same legislative history which prompted our holding in *Pacific Gas & Electric*, coupled with an examination of Congress' actions with respect to other portions of the Atomic Energy Act, convinces us that the pre-empted field does not extend as far as Kerr-McGee would have it.

As we recounted in Pacific Gas & Electric, "[u]ntil 1954 . . . the use, control, and ownership of nuclear technology remained a federal monopoly." In that year, Congress enacted legislation which provided for private involvement in the development of atomic energy. However, the Federal Government retained extensive control over the manner in which this development occurred. In particular, the Atomic Energy Commission was given "exclusive jurisdiction to license the transfer, delivery, receipt, acquisition, possession, and use of nuclear materials."

Nuclear Power

In 1959 Congress amended the Atomic Energy Act in order to "clarify the respective responsibilities . . . of the States and the Commission with respect to the regulation of byproduct, source, and special nuclear materials." The Commission was authorized to turn some of its regulatory authority over to any State which would adopt a suitable regulatory program. However, the Commission was to retain exclusive regulatory authority over "the disposal of such . . . byproduct, source, or special nuclear material as the Commission determines . . . should, because of the hazards or potential hazards thereof, not be disposed of without a license from the Commission." The States were therefore still precluded from regulating the safety aspects of these hazardous materials.

Congress' decision to prohibit the States from regulating the safety aspects of nuclear development was premised on its belief that the Commission was more qualified to determine what type of safety standards should be enacted in this complex area. As Congress was informed by the AEC, the 1959 legislation provided for continued federal control over the more hazardous materials because "the technical safety considerations are of such complexity that it is not likely that any State would be prepared to deal with them during the foreseeable future." If there were nothing more, this concern over the States' inability to formulate effective standards and the foreclosure of the States from conditioning the operation of nuclear plants on compliance with state-imposed safety standards arguably would disallow resort to state-law remedies by those suffering injuries from radiation in a nuclear plant. There is, however, ample evidence that Congress had no intention of forbidding the States to provide such remedies.

Indeed, there is no indication that Congress even seriously considered precluding the use of such remedies either when it enacted the Atomic Energy Act in 1954 or when it amended it in 1959. This silence takes on added significance in light of Congress' failure to provide any federal remedy for persons injured by such conduct. It is difficult to believe that Congress would, without comment, remove all means of judicial recourse for those injured by illegal conduct.

More importantly, the only congressional discussion concerning the relationship between the Atomic Energy Act and state tort remedies indicates that Congress assumed that such remedies would be available. After the 1954 law was enacted, private companies contemplating entry into the nuclear industry expressed concern over potentially bankrupting state-law suits arising out of a nuclear incident. As a result, in 1957 Congress passed the Price-Anderson Act. That Act established an indemnification scheme under which operators of licensed nuclear facilities could be required to obtain up to $60 million in private financial protection against such suits. The Government would then provide indemnification for the next $500 million of liability, and the resulting $560 million would be the limit of liability for any one nuclear incident.

Although the Price-Anderson Act does not apply to the present situation, the discussion preceding its enactment and subsequent amendment indicates that Congress assumed that persons injured by nuclear accidents were free to utilize existing state tort law remedies. The Joint Committee Report on the original version of the Price-Anderson Act explained the relationship between the Act and existing state tort law as follows:

> *"Since the rights of third parties who are injured are established by State law, there is no interference with the State law until there is a likelihood that the damages exceed the amount of financial responsibility required together with the amount of the indemnity. At that point the Federal interference is limited to the prohibition of making payments through the State courts and to prorating the proceeds available."*

Congress clearly began working on the Price-Anderson legislation with the assumption that in the absence of some subsequent legislative action, state tort law would apply. This was true even though Congress was fully aware of the Commission's exclusive regulatory authority over safety matters. As the Joint Committee explained in 1965:

> *"The Price-Anderson Act also contained provisions to improve the AEC's procedures for regulating reactor licensees. . . . This manifested the continuing concern of the Joint Committee and Congress with the necessity for assuring the effectiveness of the national regulatory program for protecting the health and safety of employees and the public against atomic energy hazards. The inclusion of these provisions . . . also reflected the intimate relationship which existed between Congress' concern for prevention of reactor accidents and the indemnity provisions of the Price-Anderson legislation."*

When it enacted the Price-Anderson Act, Congress was well aware of the need for effective national safety regulation. In fact, it intended to encourage such regulation. But, at the same time, "the right of the State courts to establish the liability of the persons involved in the normal way (was) maintained."

The belief that the NRC's exclusive authority to set safety standards did not foreclose the use of state tort remedies was reaffirmed when the Price-Anderson Act was amended in 1966. The 1966 amendment was designed to respond to concerns about the adequacy of state-law remedies. It provided that in the event of an "extraordinary nuclear occurrence," licensees could be required to waive any issue of fault, any charitable or governmental immunity defense, and any statute of limitations defense of less than 10 years.

Again, however, the importance of the legislation for present purposes is not so much in its substance, as in the assumptions on which it was based.

Describing the effect of the 1966 amendment, the Joint Committee stated:

> *"By requiring potential defendants to agree to waive defenses the defendants' rights are restricted; concomitantly, to this extent, the rights of plaintiffs are enlarged. Just as the rights of persons who are injured are established by State law, the rights of defendants against whom liability is asserted are fixed by State law. What this subsection does is to authorize the [NRC] to require that defendants covered by financial protection and indemnity give up some of the rights they might otherwise assert."*

Similarly, when the Committee outlined the rights of those injured in nuclear incidents which were not extraordinary nuclear occurrences, its reference point was again state law. "Absent . . . a determination [that the incident is an "extraordinary nuclear occurrence"], a claimant would have exactly the same rights that he has today under existing law—including, perhaps, benefit of a rule of strict liability if applicable State law so provides." Indeed, the entire discussion surrounding the 1966 amendment was premised on the assumption that state remedies were available notwithstanding the NRC's exclusive regulatory authority. For example, the Committee rejected a suggestion that it adopt a federal tort to replace existing state remedies, noting that such displacement of state remedies would engender great opposition. Hearings before the Joint Committee on Atomic Energy on Proposed Amendments to Price-Anderson Act Relating to Waiver of Defenses. If other provisions of the Atomic Energy Act already precluded the States from providing remedies to its citizens, there would have been no need for such concerns. Other comments made throughout the discussion were similarly based on the assumption that state remedies were available.

Kerr-McGee focuses on the differences between compensatory and punitive damages awards and asserts that, at most, Congress intended to allow the former. This argument, however, is misdirected because our inquiry is not whether Congress expressly allowed punitive damages awards. Punitive damages have long been a part of traditional state tort law. As we noted above, Congress assumed that traditional principles of state tort law would apply with full force unless they were expressly supplanted. Thus, it is Kerr-McGee's burden to show that Congress intended to preclude such awards. Yet, the company is unable to point to anything in the legislative history or in the regulations that indicates that punitive damages were not to be allowed.

Appendix E

In sum, it is clear that in enacting and amending the Price-Anderson Act, Congress assumed that state-law remedies, in whatever form they might take, were available to those injured by nuclear incidents. This was so even though it was well aware of the NRC's exclusive authority to regulate safety matters. No doubt there is tension between the conclusion that safety regulation is the exclusive concern of the federal law and the conclusion that a State may nevertheless award damages based on its own law of liability. But as we understand what was done over the years in the legislation concerning nuclear energy, Congress intended to stand by both concepts and to tolerate whatever tension there was between them. We can do no less. It may be that the award of damages based on the state law of negligence or strict liability is regulatory in the sense that a nuclear plant will be threatened with damages liability if it does not conform to state standards, but that regulatory consequence was something that Congress was quite willing to accept.

We do not suggest that there could never be an instance in which the federal law would pre-empt the recovery of damages based on state law. But insofar as damages for radiation injuries are concerned, pre-emption should not be judged on the basis that the Federal Government has so completely occupied the field of safety that state remedies are foreclosed but on whether there is an irreconcilable conflict between the federal and state standards or whether the imposition of a state standard in a damages action would frustrate the objectives of the federal law. We perceive no such conflict or frustration in the circumstances of this case.

B

The United States, as amicus curiae, contends that the award of punitive damages in this case is pre-empted because it conflicts with the federal remedial scheme, noting that the NRC is authorized to impose civil penalties on licensees when federal standards have been violated. However, the award of punitive damages in the present case does not conflict with that scheme. Paying both federal fines and state-imposed punitive damages for the same incident would not appear to be physically impossible. Nor does exposure to punitive damages frustrate any purpose of the federal remedial scheme.

Kerr-McGee contends that the award is pre-empted because it frustrates Congress' express desire "to encourage widespread participation in the development and utilization of atomic energy for peaceful purposes." In *Pacific Gas & Electric*, we observed that "[t]here is little doubt that a primary purpose of the Atomic Energy Act was, and continues to be, the promotion of nuclear power." However, we also observed that "the promotion of nuclear power is not to be accomplished 'at all costs.'" Indeed, the provision cited by Kerr-McGee goes on to state that atomic energy should be developed

269

and utilized only to the extent it is consistent "with the health and safety of the public." Congress therefore disclaimed any interest in promoting the development and utilization of atomic energy by means that fail to provide adequate remedies for those who are injured by exposure to hazardous nuclear materials. Thus, the award of punitive damages in this case does not hinder the accomplishment of the purpose stated in 2013(d).

We also reject Kerr-McGee's submission that the punitive damages award in this case conflicts with Congress' express intent to preclude dual regulation of radiation hazards. As we explained in Part A, Congress did not believe that it was inconsistent to vest the NRC with exclusive regulatory authority over the safety aspects of nuclear development while at the same time allowing plaintiffs like Mr. Silkwood to recover for injuries caused by nuclear hazards. We are not authorized to second-guess that conclusion.

IV

We conclude that the award of punitive damages in this case is not preempted by federal law. . . .

The judgment of the Court of Appeals with respect to punitive damages is therefore reversed, and the case is remanded to that court for proceedings consistent with this opinion.

APPENDIX F

STATISTICS AND DATA ON NUCLEAR POWER

NUCLEAR POWER PRODUCTION IN THE UNITED STATES, 1973–2004

Year	Net Generation of Electricity (million kilowatt hours)	Nuclear Share of Electricity
1973	83,479	4.5
1974	113,976	6.1
1975	172,506	9.0
1976	191,104	9.4
1977	250,883	11.8
1978	276,403	12.5
1979	255,155	11.3
1980	251,116	11.0
1981	272,674	11.9
1982	282,773	12.6
1983	293,677	12.7
1984	327,634	13.5
1985	383,691	15.5
1986	414,038	16.6
1987	455,270	17.7
1988	526,973	19.5
1989	529,355	17.8
1990	576,862	19.0
1991	612,565	19.9
1992	618,776	20.1
1993	610,291	19.1
1994	640,440	19.7
1995	673,402	20.1
1996	674,729	19.6
1997	628,644	18.0
1998	673,702	18.6
1999	728,254	19.7
2000	753,893	19.8
2001	768,826	20.6
2002	780,064	20.2
2003	763,725	19.8
2004	788,556	19.9

Source: Monthly Energy Review. Energy Information Administration, January 2005, Table 8.1: Nuclear Energy Overview. URL: http://www.eia.doe.gov/emeu/mer/pdf/pages/sec8_3.pdf.

DESIGN OF TWO COMMON TYPES OF NUCLEAR POWER PLANTS

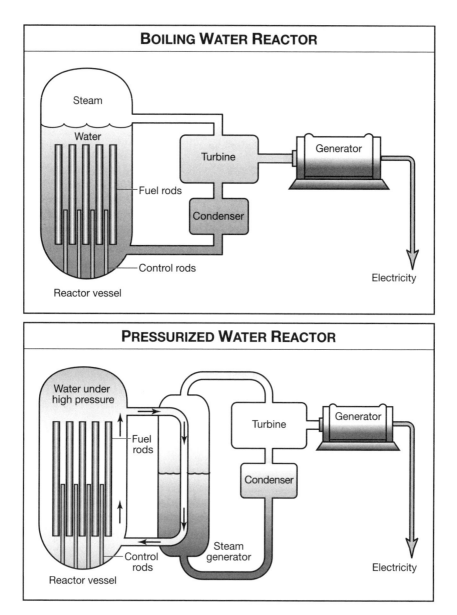

BOILING WATER REACTOR

Steam

Water

Fuel rods

Turbine

Generator

Condenser

Control rods

Electricity

Reactor vessel

PRESSURIZED WATER REACTOR

Water under high pressure

Fuel rods

Turbine

Generator

Condenser

Control rods

Steam generator

Electricity

Reactor vessel

Appendix F

SHARE OF ELECTRICITY OBTAINED FROM NUCLEAR POWER IN 2003

Country	Nuclear Electrical Power (terawatt hours)	Percentage of All Electricity from Nuclear Power
Argentina	7.0	8.6
Armenia	1.8	35.5
Belgium	44.6	55.0
Brazil	13.3	3.6
Bulgaria	16.0	37.7
Canada	70.3	12.5
China (mainland)	41.5	2.2
China (Taiwan)	37.4	21.5
Czech Republic	25.9	31.1
Finland	21.8	27.3
France	420.7	77.7
Germany	157.4	28.1
Hungary	11.0	32.7
India	16.4	3.3
Japan	230.8	25.0

Total U.S. Commercial Spent Nuclear Fuel Discharges, 1968–2002

Reactor Type	Number of Fuel Assemblies		
	Stored at Reactor Site	Stored away from Reactor Site	Total
Boiling-water	90,398	2,957	93,355
Pressurized-water	69,800	491	70,921
High-temperature gas-cooled	1,464	744	2,208
TOTAL	161,662	4,192	165,854
	Metric Tonnes of Uranium		
Boiling-water	16,153.6	554.0	16,707.6
Pressurized-water	30,099.0	192.6	30,291.6
High-temperature gas-cooled	15.4	8.8	24.2
TOTAL	46,268.0	755.4	47,023.4

Source: "Spent Nuclear Fuel Data." Energy Information Administration, URL: http://www.eia. doe.gov/cneaf/nuclear/spent_fuel/ussnfdata.html#table1, October 2004.

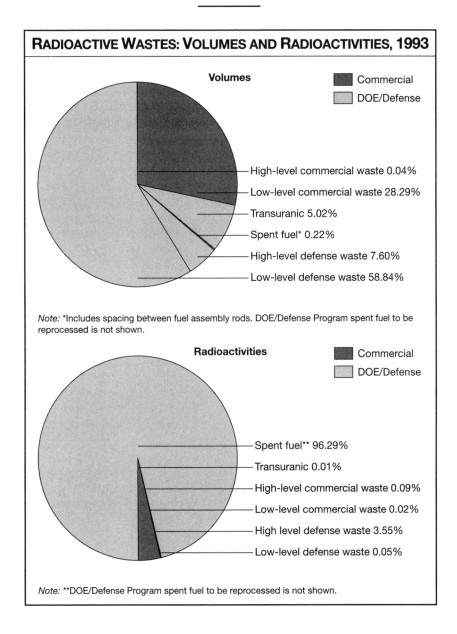

RADIOACTIVE WASTES: VOLUMES AND RADIOACTIVITIES, 1993

Volumes

Commercial

DOE/Defense

High-level commercial waste 0.04%

Low-level commercial waste 28.29%

Transuranic 5.02%

Spent fuel* 0.22%

High-level defense waste 7.60%

Low-level defense waste 58.84%

Note: *Includes spacing between fuel assembly rods. DOE/Defense Program spent fuel to be reprocessed is not shown.

Radioactivities

Commercial

DOE/Defense

Spent fuel** 96.29%

Transuranic 0.01%

High-level commercial waste 0.09%

Low-level commercial waste 0.02%

High level defense waste 3.55%

Low-level defense waste 0.05%

Note: **DOE/Defense Program spent fuel to be reprocessed is not shown.

SPENT NUCLEAR FUEL
STORAGE SITES

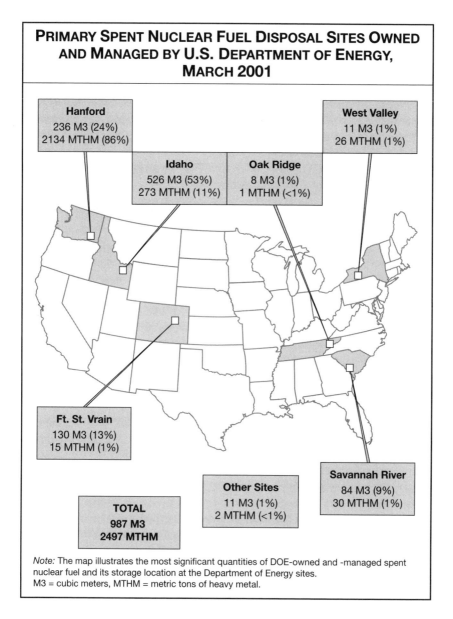

PRIMARY SPENT NUCLEAR FUEL DISPOSAL SITES OWNED AND MANAGED BY U.S. DEPARTMENT OF ENERGY, MARCH 2001

Hanford
236 M3 (24%)
2134 MTHM (86%)

West Valley
11 M3 (1%)
26 MTHM (1%)

Idaho
526 M3 (53%)
273 MTHM (11%)

Oak Ridge
8 M3 (1%)
1 MTHM (<1%)

Ft. St. Vrain
130 M3 (13%)
15 MTHM (1%)

Savannah River
84 M3 (9%)
30 MTHM (1%)

Other Sites
11 M3 (1%)
2 MTHM (<1%)

TOTAL
987 M3
2497 MTHM

Note: The map illustrates the most significant quantities of DOE-owned and -managed spent nuclear fuel and its storage location at the Department of Energy sites.
M3 = cubic meters, MTHM = metric tons of heavy metal.

NATIONS BELONGING TO THE NUCLEAR CLUB

Nation	First Fission Test	First Fusion Test
United States	July 16, 1945	October 31, 1952 (GMT) November 1, 1952 (local)
Soviet Union/Russia	August 29, 1949	August 12, 1953
Great Britain	October 3, 1952	May 15, 1957
France	February 13, 1960	August 24, 1968
China	October 16, 1964	June 17, 1967
India	May 18, 1974	May 11, 1998
Pakistan	May 28, 1998	(none tested)

Note: Other nations claim to have or are believed to have nuclear bombs although no tests have been detected for other than the above nations.

NUCLEAR POWER PLANTS CONSTRUCTED UNDER THE POWER DEMONSTRATION REACTOR PROGRAM, 1957–1962

Plant	Electrical Power Output (kilowatts)	Start-up Date
Shippingport Atomic Power Station, Shippingport, PA	60,000	1957
Dresden Nuclear Power Station, Morris, IL	208,000	1959
Yankee Nuclear Power Station, Rowe, MA	161,000	1960
Indian Point Unit No. 1, Indian Point, NY	255,000	1962
Hallam Nuclear Power Facility, Hallam, NE	75,000	1962
Big Rock Nuclear Power Plant, Big Rock Point, MI	47,800	1962
Elk River Reactor, Elk River, MN	20,000	1962

COMPONENTS OF HIGH-LEVEL NUCLEAR WASTES

Isotope	Half-life (in yrs)	Radioactivity Remaining (in Curies) After				
		10 yrs	500 yrs	1,000 yrs	10,000 yrs	100,000 yrs
Strontium-90	28	2×10^6	15	trace	none	none
Cesium-137	30	3×10^6	40	trace	none	none
Plutonium-239	24,110	2.2×10^4	2.7×10^4	2.2×10^4	5.6×10^4	8×10^3
Plutonium-240	6,540	4.9×10^4	1.75×10^5	1.7×10^5	6.8×10^4	7
Curium-245	85,000	5.6×10^4	5.2×10^4	5.2×10^4	2.5×10^4	0.5

Note: The curie is the unit used to measure the intensity of radiation produced by a material. It is equal to 37 billion disintegrations per second.

Source: James C. Warf and Sheldon C. Plotkin, "Disposal of High-Level Nuclear Waste," WagingPeace.org, September 1996, available online at http://www.wagingpeace.org/articles/1996/09/00_warf_disposal_print.htm.

277

COMPOSITION OF TEN LOW-LEVEL WASTE COMPACTS, AS OF 2004

Compact	Member States	Facility	Status
Appalachian	Delaware, Maryland, Pennsylvania, West Virginia	None Host: Pennsylvania	Exporting wastes to Utah and South Carolina
Atlantic	Connecticut, New Jersey, South Carolina	Host: Barnwell, South Carolina	Effective 2008: Will accept wastes from Atlantic Compact states only
Central	Arkansas, Kansas, Louisiana, Nebraska, Oklahoma	None Host: Nebraska	Nebraska contesting its selection as host state
Central Midwest	Kentucky, Illinois	None Host: Illinois	Exporting wastes to Utah and South Carolina
Midwest	Indiana, Iowa, Missouri, Minnesota, Ohio, Wisconsin	None Host: Ohio (site development discontinued in 1997)	Exporting wastes to South Carolina
Northwest	Alaska, Hawaii, Idaho, Montana, Utah, Washington, Wyoming	Host: Hanford, Washington	Accepting wastes from Northwest and Rocky Mountain Compacts
Rocky Mountain	Colorado, Nevada, New Mexico Host: None	None	Exporting wastes to Washington
Southeast	Alabama, Florida, Georgia, Mississippi, Tennessee, Virginia	None Host: North Carolina (in dispute)	Compact states are suing North Carolina for withdrawing its agreement to serve as host state
Southwestern	Arizona, California, North Dakota, South Dakota	None Host: California	Development of Ward Valley, CA, disposal site halted in 1999
Texas	Maine, Texas, Vermont	None Host: Texas	Site chosen in southwest Texas denied permit in 1998 by Texas Natural Resource Conservation Commission

Source: "Low-level Nuclear Waste Disposal Update (10-4-02)." URL: http://www.agiweb.org/gap/legis107/lowlevel_waste.html#compacts

Appendix F

NUCLEAR POWER PLANTS BEING DECOMMISSIONED, AS OF 2005

Reactor	Type	Location	Shut-down Date	Status
GE VBWR	BWR	Alameda Co., CA	12/9/63	SAFSTOR
CVTR	Pressure Tube, Heavy Water	Parr, SC	1/67	License Terminated
Pathfinder NRC	Superheat BWR	Sioux Fall, SD	9/16/67	DECON
Saxton	PWR	Saxton, PA	5/72	DECON
Fermi I	Fast Breeder	Monroe Co., MI	9/22/72	SAFSTOR
Indian Point I	PWR	Buchanan, NY	10/31/74	SAFSTOR
Peach Bottom I	HTGR	York Co., PA	10/31/74	SAFSTOR
Humboldt Bay 3	BWR	Eureka, IL	10/31/78	SAFSTOR
Dresden	BWR	Morris, IL	10/31/78	SAFSTOR
Three Mile Isaland 2	PWR	Middletown, PA	3/28/79	SAFSTOR
LaCrosse	BWR	LaCrosse, WI	4/30/87	SAFSTOR
Rancho Seco	PWR	Sacramento, CA	6/7/89	DECON
Shoreham	BWR	Suffolk Co., NY	6/28/89	License Terminated
Fort St. Vrain	HTGR	Platteville, CO	8/18/89	License Terminated
Yankee Rowe	PWR	Franklin Co., MA	10/1/91	DECON
Trojan	PWR	Portland, OR	11/9/92	DECON
San Onofre I	PWR	San Clemente, CA	11/30/92	DECON
Millstone	BWR	Waterford, CT	11/4/95	DECON
Haddam Neck	PWR	Haddam Neck, CT	7/22/96	DECON
Maine Yankee	PWR	Bath, ME	12/96	DECON
Big Rock Point	BWR	Charlevoix, MI	8/97	DECON
Zion I	PWR	Zion, IL	2/98	SAFSTOR
Zion 2	PWR	Zion, IL	2/98	SAFSTOR

Note: Under types of nuclear power plant, BWR stands for boiling water reactor, PWR stands for pressurized water reactor, and HTGR stands for high temperature gas cooled reactor.

Source: Nuclear Regulatory Commission. Fact Sheet on Decommissioning Nuclear Power Plants. URL: http://www.nrc.gov/reading-rm/doc-collections/fact-sheets/decommissioning.html.

279

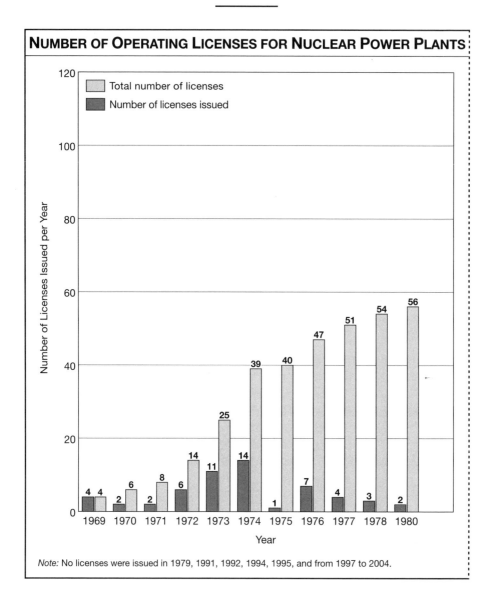

NUMBER OF OPERATING LICENSES FOR NUCLEAR POWER PLANTS

Number of Licenses Issued per Year

- Total number of licenses
- Number of licenses issued

Year

Note: No licenses were issued in 1979, 1991, 1992, 1994, 1995, and from 1997 to 2004.

Appendix F

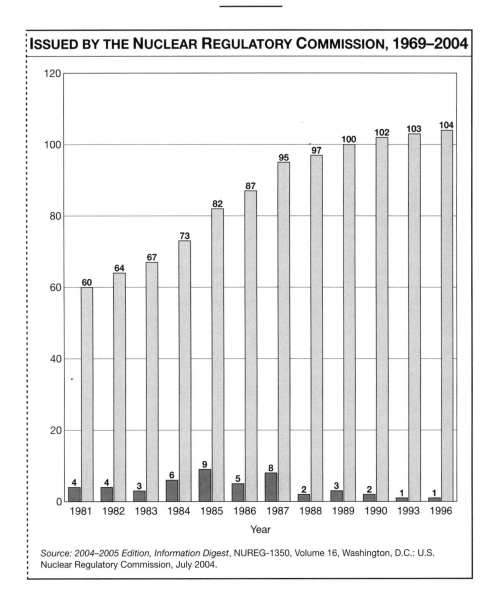

ISSUED BY THE NUCLEAR REGULATORY COMMISSION, 1969–2004

Source: 2004–2005 Edition, Information Digest, NUREG-1350, Volume 16, Washington, D.C.: U.S. Nuclear Regulatory Commission, July 2004.

281

Nuclear Power

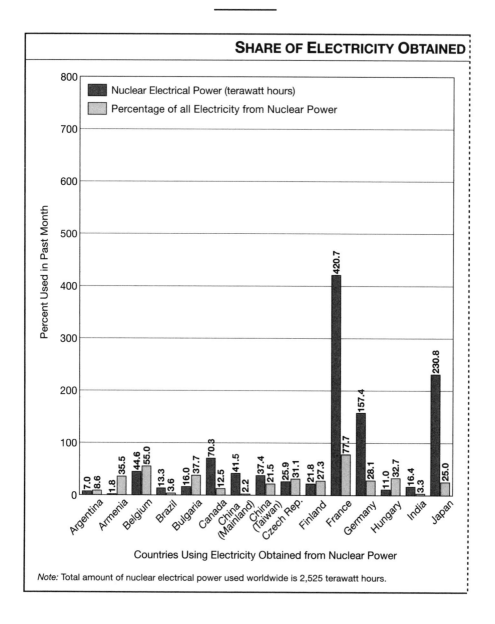

SHARE OF ELECTRICITY OBTAINED

- ■ Nuclear Electrical Power (terawatt hours)
- □ Percentage of all Electricity from Nuclear Power

Percent Used in Past Month

Countries Using Electricity Obtained from Nuclear Power

Note: Total amount of nuclear electrical power used worldwide is 2,525 terawatt hours.

Appendix F

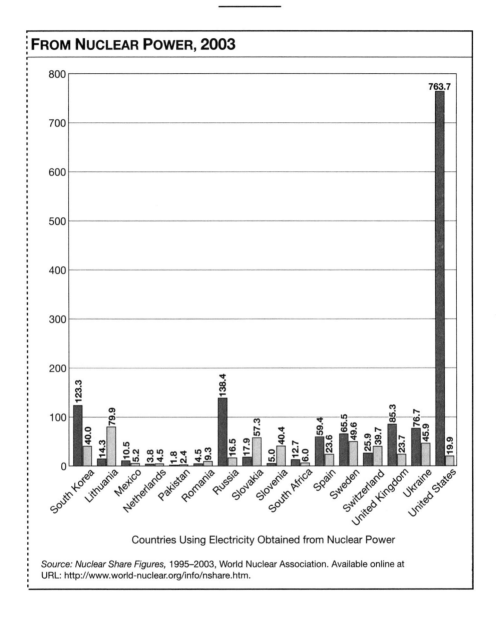

INDEX

Locators in **boldface** indicate main topics. Locators followed by *c* indicate chronology entries. Locators followed by *b* indicate biographical entries. Locators followed by *g* indicate glossary entries. Locators followed by *t* indicate graphs, tables, or diagrams.

A

AASEC. *See* Abalone Alliance Safe Energy Clearinghouse (AASEC)
Abalone Alliance 39, 95*c*, 226–227
Abalone Alliance Safe Energy Clearinghouse (AASEC) 226–227
About.com (search engine) 141
Abraham, Spencer 84
accidents at nuclear power plants 32–35
ACRS. *See* Advisory Committee on Reactor Safeguards (ACRS)
Advanced Boiling Water Reactor 100*c*
Advisory Committee on Reactor Safeguards 29, 90*c*
AEC. *See* Atomic Energy Commission
Aerojet General 20
age of fossil fuels 17
agreement state 124*g*
AIP. *See* American Institute of Physics (AIP)
aircraft carrier, nuclear. *See* nuclear aircraft carrier

Aircraft Nuclear Propulsion (ANP) project 15
airplane, nuclear-powered. *See* nuclear-powered airplane
alchemists 6
Allen, Irene H. 80
Allen v. United States **79–83**
    legal issues 80–81
    decision 81
    impact 81–82
Alliance for Nuclear Accountability (ANA) 227
AllTheWeb (search engine) 141
alpha particle 15, 85*c*, 124*g. See also* alpha rays
alpha rays 128*c*, 129*c*, *See also* alpha particles
al-Qaeda 52
American Institute of Physics (AIP) 219–220
American Nuclear Energy Council 99*c*, 225
American Nuclear Society (ANS) 220
American Physical Society 221
    review of Rasmussen Report 32

ANA. *See* Alliance for Nuclear Accountability (ANA)
Anderson, Clinton 27, 65, 106*b*, 116
ANL-W. *See* Argonne National Laboratory-West (ANL-W)
ANP. *See* Aircraft Nuclear Propulsion (ANP) project
ANS. *See* American Nuclear Society (ANS)
antinuclear movements 36–43
Arco, Idaho 19, 20, 89*c*, 90*c*
Argonne National Laboratory 19, 123
Argonne National Laboratory-West (ANL-W) 217
Arizona (nuclear power initiative) 41, 80
Arkansas Nuclear One (power plant) 102*c*
Armenia 28
arms race 11–12, 36
articles about nuclear power. *See* nuclear power, articles

284

Index

Index

287

Index

291

Nuclear Power

MHB Technical Associates 114
Michigan 46
Midland nuclear power plant (Michigan) 42, 99c
Midwest Compact 46
Midwest No-Nukes Conference 38
Military Toxics Project (MTP) 229
mill tailings 43, 67, 127g
Minor, Gregory 114–115b
Missile Envy (book) 107
Missouri (nuclear power initiative) 42
moderator 124, 125, 127–128g, 129
Montana (nuclear power initiative) 41
MTP. *See* Military Toxics Project (MTP)

N

Nader, Ralph 40, 107, 115b
Nagasaki, Japan 3, 11, 34, 36, 88c
NAM. *See* National Atomic Museum (NAM)
National Academy of Sciences (NAS) 48, 83, 90c, 108
National Association of Science Writers 13
National Atomic Museum (NAM) 224
National Council on Radiation Protection and Measurements (NCRP) 214–215
National Defense Research Committee 107
National Energy Policy 57, 102c
National Environmental Protection Act (NEPA) of 1970 69–71, 76

National Laboratory System, U.S. Department of Energy (DOE) 216–217. *See also* U.S. Department of Energy
National Nuclear Data Center (NNDC) 215
National Nuclear Security Administration (NNSA) 215
National Press Club 54
National Radiological Protection Board (United Kingdom) 101c
National Reactor Testing Station (NRTS)19–20, 34, 89c
1961 accident 92c
National Resources Defense Council (NRDC) 71-73, 229
Nautilus (submarine). *See* USS *Nautilus* (nuclear submarine)
NA-YGN. *See* North American Young Generation in Nuclear (NA-YGN)
NCI. *See* Nuclear Control Institute (NCI)
NCRP. *See* National Council on Radiation Protection and Measurements (NCRP)
NEA. *See* Nuclear Energy Agency (NEA)
NEI. *See* Nuclear Energy Institute (NEI)
NEIS. *See* Nuclear Energy Information Service (NEIS)
NEPA. *See* Nuclear Energy Propulsion Aircraft project
neutrino 128g
neutron 5, 128g
Nevada 46

Nevada Test Site 79
New Hampshire 46
New Nuclear Danger, The (book) 107
New York 46
New York Times 78, 95c, 111, 118
n,γ reaction 6–7
NIPSCO. *See* Northern Indiana Public Service Company (NIPSCO)
NIRS/WISE. *See* Nuclear Information and Resource Service & World Information Service on Energy (NIRS/WISE)
NNDC. *See* National Nuclear Data Center (NNDC)
NNSA. *See* National Nuclear Security Administration (NNSA)
Norris, George 115b
North American Young Generation in Nuclear (NA-YGN) 224–225
North Carolina 46
Northern Indiana Public Service Company (NIPSCO) 38
Northern Indiana Public Service Co. v. Porter County Chapter of the Izaak Walton League of America, Inc., et al. (423 US 12) 38
NPRI. *See* Nuclear Power Research Institute (NPRI)
NRC. *See* Nuclear Regulatory Commission (NRC)
NRDC. *See* National Resources Defense Council (NRDC)

292

Index